NONLINEAR SYSTEMS

VOLUME II, APPLICATIONS to BILINEAR CONTROL

Ronald R. Mohler
Oregon State University

PRENTICE HALL, Englewood Cliffs, New Jersey 07632

Library of Congress Cataloging-in-Publication Data

Mohler, Ronald R.
Nonlinear systems / Ronald R. Mohler.
p. cm.
Includes bibliographical references.
Contents: v. 1. Dynamics and control -- v. 2. Applications to bilinear control.
ISBN 0-13-623489-5 (v. 1). -- ISBN 0-13-623521-2 (v. 2)
1. Control theory. 2. Nonlinear theories. I. Title.
QA402.3.M634 1991
003'.75--dc20 89-26633
CIP

Editorial/production supervision and
interior design: PATRICE FRACCIO
Cover design: EDSAL ENTERPRISES
Manufacturing buyer: LORI BULWIN

Printed in the United States of America
10 9 8 7 6 5 4 3 2 1

ISBN 0-13-623521-2

Prentice-Hall International (UK) Limited, *London*
Prentice-Hall of Australia Pty. Limited, *Sydney*
Prentice-Hall Canada Inc., *Toronto*
Prentice-Hall Hispanoamericana, S.A., *Mexico*
Prentice-Hall of India Private Limited, *New Delhi*
Prentice-Hall of Japan, Inc., *Tokyo*
Simon & Schuster Asia Pte. Ltd, *Singapore*
Editora Prentice-Hall do Brasil, Ltda., *Rio de Janeiro*

To Curtis, John, and Andrew

CONTENTS

PREFACE

Volume 2 of *Nonlinear Systems* is a logical continuation of Vol. 1, Dynamics and Control. Originally, I had intended to present the combined material in one integrated text. The single volume, unfortunately, became a bit unwieldy. Consequently, Vol. 1 evolved as an applications-oriented, basic nonlinear dynamics and control text which suits the role of a traditional engineering or applied-mathematics course. Vol. 2, Applications to Bilinear Control, is more specialized. It is intended for a special-topics course in systems and control, applied mathematics, or biomedicine. Like Vol. 1, this book is flavored with multidisciplinary applications. I hope that it will be useful for practical applications as well as continued research.

The material, which is presented here, has been taught on a regular basis in a single-term graduate course in electrical and computer engineering at Oregon State University and on a visiting basis at the Naval Postgraduate School. Invited sequences of research-oriented lectures on this material were presented at the University of Rome, the Soviet Academy of Sciences, the University of California at Los Angeles, the University of London, the Australian National University, the University of Lund, the University of Manchester, the Chinese Academy of Sciences, and the International Institute for Applied Systems Analysis. Over one hundred seminars on related research have been presented at universities, research centers, and industries throughout the world. The numerous discussions with colleagues everywhere have been most gratifying and have played a major role in this book.

After the introduction in Chapter 1, typical applications are summarized in Chapter 2. Since engineering applications were emphasized in Vol. 1, biomedical processes and, particularly, mathematical models in immunology and disease control are studied here. Then, by means of Liapunov's direct method, studied in Vol. 1, the stability of bilinear systems is investigated in Chapter 3. Structural properties, related to controllability, observer design, and connections of bilinear systems are presented in Chapter 4. Here a successful observer is designed for a field-controlled dc motor with an inertial load.

Except for certain classes, such as weakly bilinear systems, nonlinear systems generally are not amenable to closed-form solution. Such solutions and approximating solutions, by means of Volterra series and Walsh functions, are studied in Chapter 5. Again, the frequency-domain approach and the transfer-function concept are convenient. Applications include the parameter identification of a model for the generation of antibodies as a control response to infective disease. Although conceptually and computationally convenient, these techniques generally are omitted from nonlinear system texts.

Chapter 6 summarizes optimal bilinear control by means of the maximum principle. This chapter forms a base for concluding design applications in Chapter 7. It is shown that nonlinear control, while presenting complications, can be used advantageously by the system designer. Here a computational algorithm is derived and demonstrated by its application to surface vehicle control. Other case studies include control of a nuclear rocket engine and a scheme for adaptive aircraft control.

The book continues in the vein of Vol. 1 to show that bilinear and related nonlinear structures arise naturally in practice. Although such complexities may be difficult to deal with, they can also be utilized to enhance system stability, controllability, and overall performance.

Although my first research on bilinear systems was sponsored by the National Science Foundation at the University of New Mexico in 1966, the concept was born out of previous research and doctoral dissertation at the Los Alamos National Laboratory and the University of Michigan. I am grateful to J. Perry, G. White, G. Pimbley, R. Howe, and E. Gilbert for this initial opportunity.

It is impossible to acknowledge the numerous contributions of all my colleagues, former professors, and students whom have made this undertaking possible. Foremost, I am grateful to A. V. Balakrishnan, G. I. Marchuk, and A. Ruberti for their professional collaboration and advice. Although virtually all of my former students have contributed indirectly, several especially noteworthy topics, which are presented here, were made by R. E. Rink, W. J. Kolodziej, C. S. Hsu, V. R. Karanam, R. Baheti, S. F. Moon, W. D. Smith, Z. Tang, T. Kwon, and F. J. Bugnon. Also, R. Brockett, C. Bruni, D. Elliott, E. Hofer, A. Isidori, G. Koch, A. Pacut, and T. Tarn, to note a few, have been very instrumental in this work. Oregon State University, Australian National University, and the International Institute for Applied Systems Analysis provided the excellent environment and support conducive to the completion of the manuscript. I am most grateful that my research on related topics has been sponsored by the NSF, ONR, EPRI, NASA, and BPA. Typing and processing of the book was completed under sometimes trying circumstances thanks to the excellent contribution of Kay Hearder, April Melton, Laurie Dockendorf, and Edith Gruber. Thanks also for the patience and understanding of my wife, Nancy.

Ron Mohler

CHAPTER 1

INTRODUCTION to BILINEAR SYSTEMS

1.1 INTRODUCTORY REMARKS

The purpose of this text is to study some of the more recent applications of nonlinear system theory to bilinear, and related variable-structure, or adaptive, control processes. A base for this work is presented in Vol. 1, *Dynamics and Control*, of this sequence on nonlinear systems [M1]. In keeping with the intent of the sequence, emphasis is given to multidisciplinary applications rather than to rigorous mathematical details. It is hoped that a cross-fertilization of ideas will evolve from a variety of applications.

As noted in Chapter 1 of Vol. 1, nonlinear, or at least bilinear, analysis may result from the very nature of the process that is to be controlled or analyzed, or the desired system performance may dictate a nonlinear dynamical design and synthesis. For example, linear small-perturbation methods and subsequent controller design may be adequate for steady-state operation of a chemical plant with small disturbances. For changing steady states, for startup and shutdown, and even for large disturbances from steady state, however, such methodologies generally are not adequate.

In Vol. 1 it is shown that bilinear systems (BLS) arise as natural models for a variety of dynamical processes. These are summarized in Chapter 2 of the present book together with certain detailed models in biomedicine. Here the immune system's control of disease is studied. As in most practical applications, the system is comprised of a connection of BLS. The coupling between the bilinear subsystems typically is nonlinear but nondynamical. Consequently, the overall closed-loop process equivalently is modeled as a bilinear system with nonlinear feedback. Or if appropriate terms appearing nonlinearly are measured, the complete synthesis is referred to as a conditionally linear system (CLS). Such CLS or BLS with state feedback represents an exact synthesis of many nonlinear systems studied in Vol. 1 (e.g., the Van der Pol equation).

BLS may be used accurately to approximate other nonlinear systems, such as by the linear and bilinear terms of the Taylor series as studied in Vol. 1. In other cases, nonlinear systems, which are not directly recognized as bilinear in form, may be transformed to BLS by an appropriate redefinition of state variables. This is studied in Chapter 4, where it is shown that BLS generally are more controllable. While asymptotically stable linear systems as well as unstable linear systems (both with magnitude bounded control) are not controllable, it is shown that judicious (adaptive) use of instability and stability can result in a controllable BLS.

Liapunov's second, or direct, method, which is developed in Vol. 1, Chapter 6, is used to study BLS stability in Chapter 3 of this book. It is shown that in some cases linear state feedback and in other cases quadratic feedback may be utilized for effective stabilization of BLS. Sometimes, BLS feedback can be used to enhance significantly the state-space region of attraction to equilibrium. In typical practical applications, classical linear feedback control theory is useful for "local" stability and control while nonlinear control is necessary "in the large." BLS solutions, Volterra series, and Walsh functions are presented in Chapter 5. Here it is shown that the Volterra series for BLS may be decomposed into an infinite number of linear-system solutions in general. In the case of weakly BLS, the Volterra series is finite and the solution results in weighted products of linear system solutions. Walsh functions are found convenient for parameter identification or for output solution approximation of BLS.

Optimal control of BLS is analyzed in Chapter 6. It is shown that BLS control may offer better performance than linear control. A second-order bilinear regulator conveniently shows the advantage of bilinear control over linear control by means of variable damping. Then, design case studies and the significance of optimization studies in practice are studied. Optimal controllers are seldom used in practice since they frequently operate on the verge of system failure (constraint violation) or since the underlying assumptions may not be satisfied. Still the analysis does provide information about the sensitivity of performance relative to different design constraints.

1.2 DEFINITIONS

Bilinear systems (BLS) comprise perhaps the simplest class of nonlinear systems (NLS). These are systems that are linear in control and linear in state but not jointly linear in state and control. In other words, they involve products of state and control. Hence it is readily seen that superposition does not apply. Consequently, they are slight generalizations of linear systems, and are a simple class of adaptive or variable-structure systems—sometimes referred to as linear systems with parametric control. Simple engineering applications include variable-impedance networks and variable-field-controlled motors. But the applications are almost endless [M1–M3].

Obviously, linear systems are a special class of BLS, but otherwise the simplest example is given by

$$\frac{dx}{dt} = ux,$$

where x is a scalar function of time and u is a scalar control function. In demography, the control u might be a coefficient of birth rate minus death rate plus migration rate into a region of population x. Time invariant bilinear systems, for our purposes, have a state form as follows:

$$\frac{d\mathbf{x}(t)}{dt} = \mathbf{Ax}(t) + \mathbf{Bu}(t) + \sum_{k=1}^{m} \mathbf{N}_k u_k(t)\mathbf{x}(t), \tag{1.1}$$

where state $\mathbf{x} \in \mathbb{R}^n$; control $\mathbf{u} \in \mathbb{R}^m$ with components $u_1, \ldots, u_m$, and $\mathbf{N}_k$, $k = 1, \ldots, m$, are $n \times n$ matrices (i.e., $\mathbf{N}_k \in \mathbb{R}^{n \times n}$). Sometimes, the last term in (1.1) is written simply as $\mathbf{N}(\mathbf{u}, \mathbf{x})$ or $\mathbf{Nu} \circ \mathbf{x}$. The corresponding BLS state diagram is shown in Fig. 1.1. The output may be generated by $\mathbf{y}(t) = \mathbf{Cx}(t)$.

In almost all applications, the additive control is independent of the multiplicative control, and the matrices $\mathbf{B}$ and $\mathbf{N}_k$, $k = 1, \ldots, m$, are so constructed. Mathematically, it is sometimes convenient to assume that $\mathbf{B}$ is a zero matrix so that (1.1) is a homogeneous BLS which theoretically may be assumed merely by addending the state with $x_{n+1} = 1$, appropriately adding extra zeros to $\mathbf{A}$ and $\mathbf{N}_k$, and appropriately including $\mathbf{B}$ in the new $\mathbf{N}_k$ [B1]. Nonlinear systems that are linear in control and state-affine systems are slight generalizations of BLS, as defined by (1.1).

In some practical systems (e.g., nuclear fission reactors, and immunology as studied in Chapter 2) bilinear systems are coupled together so that the total system is more highly nonlinear. Such system coupling may be linear or nonlinear, depending on the particular process under investigation. For example, the output of one subsystem, say reactor power, may become an input to another subsystem, such as heat transfer. In the nuclear power reactor, which is studied briefly in Vol. 1, Chapter 2, appropriate temperature from the heat exchange becomes a feedback variable via

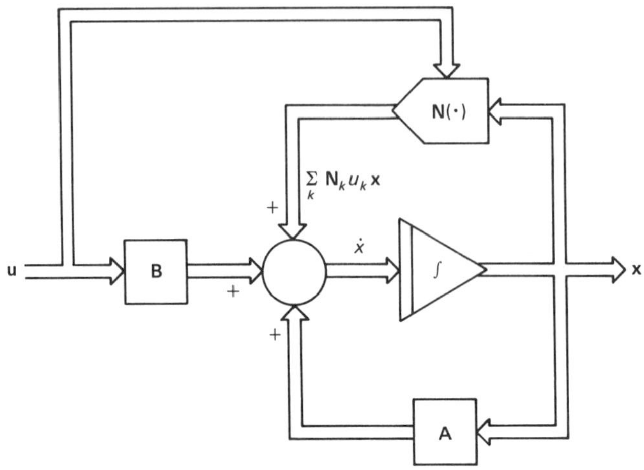

Figure 1.1 Bilinear system state diagram.

temperature reactivity (or neutron multiplication) to the nuclear fission process which generates power. Such systems are called coupled bilinear systems here.

An interesting class of coupled bilinear systems results if the nonlinearly appearing states are measured. This case may be interpreted as a BLS with measurable nonlinear feedback terms generating the closed-loop controls. In other words, with a scalar control $u(\cdot)$ a function of measured output $\mathbf{y}(t)$, (1.1) takes the following form:

$$\frac{d\mathbf{x}(t)}{dt} = [\mathbf{A} + \mathbf{N}(\mathbf{y}(t))]\mathbf{x}(t) + \mathbf{g}(\mathbf{y}(t)), \tag{1.2}$$

and the output $\mathbf{y}(t)$ is composed of only those components of $\mathbf{x}(t)$ that are measured. Since for a given set of measurements $\mathbf{y}(t)$, the system is linear (i.e., conditionally linear), (1.2) is called a conditionally linear system. Equation (1.2) can also be interpreted as a model that is linear in nonmeasurable states but nonlinear in the measurable states. It seems to be a reasonable way to approximate a nonlinear process with little known information about structure since the nonlinearities are based on measured information. Such measurements may detect a deviation from linearity and make an appropriate model change in structure. These conditionally linear systems will be studied further in subsequent chapters. Finite-state estimators (nonlinear filters) have been derived for stochastic representations of such systems. The interested reader is referred to [K1] and [M4] for more details. Similar to linear systems and bilinear systems, deterministic state observers may be generated for the unmeasured states of the conditionally linear system for purposes of feedback controller design.

1.3 HISTORICAL DEVELOPMENT

I first realized in the early 1960s that BLS have certain interesting properties and are quite prevalent in nature. In this sense, BLS grew out of research on nuclear reactor control and nuclear rocket propulsion at the Los Alamos National Laboratory [M5]–[M8]. Here nuclear fission, convective heat-transfer, and turbo-pump dynamics may all be modeled approximately by BLS. In the first case, neutron multiplication or reactivity is a control with neutrons and precursors as states. With heat-transfer coefficient or coolant mass flow rate as a control, and temperatures as states, and with pump speed as a state and valve position a control, the latter two yield approximate BLS [M1]. Note, however, that while the open-loop processes are BLS, the total system involves more highly nonlinear systems or coupled BLS.

In 1966, NSF awarded us the first grant on bilinear-system research at the University of New Mexico. Even much earlier, however, interest in systems linear in control had evolved with regard to reachable zones and optimization by Caratheodory [C1], who introduced Pfaffian equations with a variety of application to thermodynamics as well as information and system theory. Later, Hermes [H1] derived controllability from integrability of the associated Pfaffian equations, and Balakrishnan [B2] showed that a large class of nonlinear systems may be modeled by systems linear in control. Still later, Krener [K2], Sussman [S1], and Fliess and Normand-Cyrot [F1] showed that under certain restrictions nonlinear input–output maps may be approximated accurately by BLS. Unfortunately, no constructive approach has been developed to make such approximations.

Meanwhile, as early as 1953, Lewis [L1] demonstrated the use of variable-gain, tachometer feedback to vary the damping of a second-order servomechanism—a simple BLS with rate feedback. In this manner, the step response is rapid and lowly damped early, and is highly damped as the output approaches steady state with little or no overshoot. Later, we analyzed the improved performance of bilinear regulators over linear regulators in more detail [M9]. Here, a simple suboptimal control is synthesized which is based on sufficient optimization conditions from Boltyanski–Bellman–Hamilton–Jacobi theory. Hofer and Tibken [H2] solved the quadratic bilinear regulator problem as a sequence of linear regulator solutions.

In 1968, we showed the lack of controllability for asymptotically stable linear systems with bounded control and derived sufficient conditions for controllability or control-constrained BLS ([R1], [M2], and [M10]). Research on BLS was given added impetus in 1969 when Ruberti developed a concentrated program at the University of Rome. Three U.S.–Italy Meetings on Variable Structure Systems (with particular emphasis to BLS) grew out of this collaboration ([M3], [R2], and [M11]). Thereafter, numerous conference sessions and journal volumes were devoted to BLS. New publications on BLS appear monthly and only a sample of the relevant contributions can be noted here.

In his analysis of controllability, Kucera [K3] first showed the role of Lie algebra for special BLS, but the technique was most fully developed by Brockett ([B3],

TABLE 2.1 APPROXIMATE BLS/CLS

Process	Multiplicative Control	Additive Control
Electrical networks	Variable impedance	Voltage/current
Mechanical links	Parametric functions of measurements, gear ratios	Torque/force
Surface vehicle	Friction brake	Engine torque or accelerator
Heat transfer	Heat-transfer coefficient (impedance)	Heat power
Nuclear fission	Reactivity or multiplication	Neutron source
Airplanes	Aerodynamic coefficient (e.g., wing flaps)	Thrust, conventional surface controls (approximately additive)
Fluid flow	Transfer rates between pools	
Socioeconomics	Tax and investment portions of profit	
Chemistry	Catalytic control, blend of agents	
Demography	Birthrate–death rate–emigration rate	
Immunology	Enzyme concentration, cell reactivity, binding coefficient	Cell source rate, inoculation rate
Agriculture	Cattle feed, fertilizer	Irrigation
Respiratory chemostat	Respiratory rate, cardiac output	Metabolic rate
Cardiovascular regulator	Flow admittances	Cardiac output
Hormone regulation	Free protein concentration	
Urban processes	Attraction multipliers	
Kidney water balance	Effective tubile permeability	
Combat models	Area fire control	
Predator–prey models	Coupled population coefficient	

2.2 PHYSICAL MODELS

Whenever parametric (coefficient) control of an otherwise linear system arises, the resulting process becomes a bilinear control system (e.g., controlled impedance networks, field-controlled electric motors, stroke-controlled internal-combustion engines, to varying degrees of accuracy). The frictional force between the automobile brakeshoe and drum is nearly proportional to the product of the orthogonal force u_1 between the surfaces and their relative velocity. Then by a summation of engine force, u_2, with inertial, braking, and other frictional forces, it is seen that the state of the vehicle is given by

$$\frac{d\mathbf{x}}{dt} = \mathbf{Ax} + u_1\mathbf{Bx} + \mathbf{c}u_2, \tag{2.1}$$

where $\mathbf{x}^{\mathrm{T}} = [x \quad \dot{x}]$; x is position;

$$\mathbf{A} = \begin{bmatrix} 0 & 1 \\ 0 & \dfrac{-kc_f}{m} \end{bmatrix}, \qquad \mathbf{B} = \begin{bmatrix} 0 & 0 \\ 0 & \dfrac{-kc_b}{m} \end{bmatrix}, \qquad \mathbf{c} = \begin{bmatrix} 0 \\ 1 \end{bmatrix};$$

k is a proportionality constant, c_f a vehicle frictional constant, c_b a brake constant, and m the vehicle mass. It is interesting that the brake pedal introduces parametric control and the accelerator, additive control. A similar phenomenon is realized in the landing mode of a high-performance aircraft. During this phase of flight the aircraft drag coefficient is controlled by means of relatively large wing "flaps." Indeed, one can feel the change in damping during this phase. While the parametric control allows improved state controllability, the additive control in terms of thrust reversers are necessary to bring the aircraft to a rapid equilibrium at the state origin. Such structural changes frequently allow systems to reach otherwise unattainable states, while the additive control is usually required for local controllability to equilibrium, as further explained by early results for BLS controllability in Chapter 4. The landing aircraft, of course, is more highly nonlinear due to altitude and Mach changes. However, the BLS approximation is much closer to reality than is a linear perturbation model. Also, it should be noted that energy dissipation in the mechanical brake more slowly stops the aircraft and indeed is the traditional practice.

Heating, ventilating, and air conditioning systems, including solar panels, storage tanks, and heat pumps, may be similarly modeled by BLS. Such analyses result in significant energy savings over that of the traditional linear optimal control approach ([R3] and [M13]). Espana and Landau [E1] analyze compartmental BLS for distillation columns with chemical concentrations as states, and flow rates, heat, and feed-flow concentrations as controls. Chemical process control in general includes many BLS approximations due to the mass-action principle, catalytic control, and heat transfer. The problem of controlling a force attitude to a flexible body, as a BLS, is studied by Slemrod [S2]. Bellini et al. [B6] derive a BLS model for a frequency-controlled induction motor. Baheti and Mohler [B7] develop a BLS model for a furnace from experimental data. Obviously, the developments can go on and on from physical principles and approximation techniques. In digital control, variable sampling rate, and in numerical analysis, variable step size or predictor slope, may be treated as bilinear controls for certain analytical purposes. Thin-film transistors, field-effect tetrodes, and field-effect transistors (JFET and MOSFET), when used as voltage-controlled resistors, may all result in bilinear network models. In communications, the phase-locked loop leads to a BLS if sine and cosine functions are state variables. Other natural BLS models that come to mind from discussions with other scientists include electromagnetic separation of isotopes (P. Kokotovich, University of Illinois), sunspot activity and earthquakes (T.S. Rao, University of Manchester, Institute of Science and Technology, U.K.), and image processing through coherent light or other transmitting medium (J. C. Dainty, Imperial College, U.K.).

2.3 BIOMEDICAL PROCESSES

Certain basic physiologic control processes that are essential to our survival are introduced here and in Vol. 1. These include regulation of carbon dioxide, blood pressure and immune response [M2], [M14]–[M17]. Compartmental structures and their relation to tracer experiments are then introduced. Although these models are simplistic representations of the real world, they are nonlinear and representative of the structure of more complex models. Mammalian temperature regulation is studied in Section 2.4, Vol. 1. Such models are studied here with the feeling that systems research and design methods can make a significant contribution to biomedicine. Again, systems engineers may have a lot to learn from nature—in this case physiology. Other biological BLS models include blood coagulation by Elliott [E2] and circulation of blood thyroxin by Distefano [D3]. In economics, BLS arise in numerous ways, such as parametric control via capital investment and tax rates. Several examples are presented in detail in [A2], [D4], [K4], and [V1].

Numerous bilinear models have been studied in ecology and agriculture. These include grassland models [P1], where additive control includes sunlight, temperature, and moisture with parametric control modeling harvest rates and grazing. Other models include generation of beef with feed rate control and pest management control by means of pesticides, pheromones, and predators as well as the control of others [M2], [L4], and [O1].

Deterministic and stochastic bilinear population models are analyzed in [M18]. In addition to human demography these models apply, for example, to general biological species, biological cells, chemical molecules, atomic particles [K5], and manufacturing and distribution of products. Similar to the cellular processes in Section 2.4, there may be birthrates, death rates, migration rates between compartments, and sometimes differentiation rates which are proportional to the population themselves. Control is introduced by manipulation of appropriate coefficients such as birthrate coefficient, cellular or neutron multiplication, and migration-attractiveness multipliers or membrane permeability.

2.3.1 Pulmocardiovascular Regulation

In this section, which follows [M2] and [G2], the respiratory system is regarded as a chemical O_2–CO_2 exchanger. Air breathed into the lungs is processed to provide oxygen to the body plant. Arterial blood, transporting oxygen to the body tissue, is in chemical equilibrium with the gas in the lungs. In the tissue, cellular metabolism converts O_2 to CO_2, which is transported back to the lungs by venous blood. The lungs, in turn, are continuously ventilated by breathing. Physiologists have long recognized that this system is self-regulating. That is, if the body tissue for some reason becomes deficient in O_2 or overloaded with CO_2, the rate of ventilation of the lungs is automatically increased, thereby tending to increase the ratio of O_2 to CO_2 in the lungs and in the arterial blood.

The model presented here was derived in [G2] to account for the transient response to stepwise CO_2 inhalation. From test data, it is evident that the ventilation lags the arterial partial pressure pCO_2 in time, and that the latter can show overshoot and undershoot. These facts suggest that it is the CO_2 concentration in body tissue that is fed back, since tissue has a large capacity for CO_2 storage. This model ignores O_2 and assumes that the respiratory center can be lumped with all other tissues into a single homogeneous CO_2 reservoir with constant blood supply [G3]. The accuracy of the model when compared to experimental data does substantiate the model for at least the purpose served here. For this model, lungs and tissue form the two compartments, with x_2 and x_1 representing the CO_2 concentrations in tissue and lung, respectively. The analogy to an *RC* circuit is rather obvious. The tissue and lung CO_2 reservoirs correspond to capacitors, while the finite flow and pressure-dependent CO_2 absorption characteristics of the blood represent a resistance to the flow of CO_2.

Again, the system of equations may be derived from continuity or conservation equations on each compartment. Beginning with the lung reservoir, the CO_2 continuity equation states that the rate of change of lung CO_2 concentration $\dot{x}_1$ is equal to the difference between the rates of CO_2 inflow and outflow by all paths divided by the volume of the reservoir V_1. Thus

$$\frac{dx_1}{dt} = \frac{1}{V_1}[uF^i(CO_2) + q_3 - q_2 - x_1 u], \tag{2.2}$$

where $F^i(CO_2)$ is the fraction of CO_2 in the inspired air, q_3 the rate at which CO_2 is delivered by venous blood, and q_2 and $x_1 u$ the rates at which CO_2 leaves the lung via arterial blood and expired air, respectively. A similar CO_2 continuity equation can be written for the tissue reservoir, so that

$$\frac{dx_2}{dt} = \frac{1}{V_2}[MR + q_2 - q_3], \tag{2.3}$$

where V_2 is tissue volume. It is assumed that the arterial blood is in chemical equilibrium with the air in the lungs. Then the arterial CO_2 concentration can be expressed as a function x_1 by means of linearized CO_2 absorption curve, as

$$\frac{q_2}{Q} = B\alpha x_1 + \beta, \tag{2.4}$$

where Q is the cardiac output, B is the barometric pressure, and α and β the slope and intercept, respectively, of the linear absorption curve. We also assume that the venous blood is in equilibrium with the tissue CO_2 concentration, and that tissue and blood have the same CO_2 absorption curve. Then we simply have

$$x_2 = \frac{q_3}{Q}. \tag{2.5}$$

Finally, using (2.4) and (2.5) to eliminate q_2 and q_3 from (2.2) and (2.3), we obtain the equations of the controlled system:

$$\frac{dx_1}{dt} = \frac{1}{V_1}[uF^i(CO_2) - x_1) + Q(x_2 - B\alpha x_1 - \beta)], \tag{2.6}$$

$$\frac{dx_2}{dt} = \frac{1}{V_2}[MR - Q(x_2 - B\alpha x_1 - \beta)]. \tag{2.7}$$

We observe that the system is bilinear. Of course, once the control law giving u as a function of state variables is substituted in the equation, the closed-loop system again becomes nonbilinear.

It was suggested that the controller receives an input signal from the brain, pCO_{2B}, and from this generates a continuously operating proportional control [G3]. It operates on the mean absolute value of the time derivative of the arterial pCO_2 and on the existing average magnitude of ventilation rate. Each of these signals is compared with a reference and multiplied by a gain constant, and the products are then added. If the sum Y is negative, it is simply added to the proportional signal from the brain CO_2 to generate the final output. If, however, Y is positive, its integral is added to the proportional brain tissue signal. It turns out that this decision process is dictated by the relationship between u and the average arterial pCO_2 time derivative magnitude. In exercise these two quantities maintain the same relative size as they have at rest. The controller thereby senses that the CO_2 load is a metabolic one and the integrator operates to zero the steady-state error in mean arterial pCO_2. In CO_2 inhalation, however, the average arterial pCO_2 time derivative magnitude is relatively small. The controller thereby senses that the CO_2 load is an external one, and the integrator is bypassed.

Control of the cardiovascular system is somewhat similar to, but considerably more complex than, the respiratory chemostat. Still, the system can be compartmented and conservation questions can be derived to describe the compartmental dynamics. Again, the ensuing model is approximately bilinear, with compartmental blood pressures as state variables and resistances to blood flow in peripheral arterioles as multiplicative controls. For example, these resistances are altered by the vasomotor nerves during exercise, hemorrhage, or shock.

The cardiovascular plant being controlled is comprised of a mechanical section (heart and vascular networks) and a chemical exchanger (tissue). The controller consists of neural and endocrine sections which are activated by error signals received through cardiac and vascular centers in the medulla in response to pressure feedbacks through arterial baroceptors. The primary controlled variables of the system are the levels of tissue (venous) pO_2 in the several parallel systemic circuits. It is assumed that other tissue agents, such as pCO_2, pH, and metabolites, are represented indirectly through pO_2.

Within the plant, the heart provides cardiac output as an input to the networks. Arterial and venous pressures are fed back to the heart, and local systemic flows feed the tissue gas exchanger. Tissue (venous) pressures, pO_2, feed back to affect systemic circuit pressures and local flows through manipulation of systemic arteriolar resistances. Network pressures are fed back to vary cardiac output. The neural signals to the heart manipulate cardiac output through heart frequency, contractile strength, and filling impedance. Neural signals, also, manipulate arteriolar resistance and venous compliance, and activate the adrenal medulla for release of epinephrine and norepinephrine. These in turn affect the heart and circuits.

The mechanical portion of the cardiovascular system may be divided into four compartments, with $v_2(t)$, cardiac output from the right heart, and $v_1(t)$, cardiac output from the left heart, as inputs to an arterial and a venous compartment, respectively [G2]. The four conservation equations are

$$c_1 \frac{dx_1}{dt} = u_1(x_2 - x_1) + v_1, \tag{2.8}$$

$$c_2 \frac{dx_2}{dt} = u_1(x_1 - x_2) - v_2, \tag{2.9}$$

$$c_3 \frac{dx_3}{dt} = u_2(x_2 - x_3) + v_2, \tag{2.10}$$

and

$$c_4 \frac{dx_4}{dt} = u_2(x_3 - x_4) - v_1, \tag{2.11}$$

where $x_1, \ldots, x_4$ are compartmental pressures in the arteries filled from the left heart, veins to the right heart, arteries filled from the right heart, and veins to the left heart, respectively; $c_1, \ldots, c_4$ are corresponding compartmental capacities (assumed constant); and $u_1(t)$ and $u_2(t)$ are inverse pulmonary-arterial and pulmonary-venous network resistances to flow. Since it is a closed system, pressures are constrained so that

$$c_1x_1 + c_2x_2 + c_3x_3 + c_4x_4 = B,$$

where B is a positive constant.

Cardiac outputs obtained from the left and right hearts are additive control functions with complicated nonlinear feedback approximated by [G2]

$$v_1 = \frac{k_4x_4}{k_1x_1 + k_5}$$

and

$$v_2 = \frac{k_2x_2}{k_3x_3 + k_5}.$$

It must be emphasized that only a very crude partial model has been studied here. More details of this model are presented in [G2].

In summary, cardiovascular regulation does involve a variable structure which may be approximated by a bilinear system as a result of multiplication of pressures by controlled resistance terms or susceptances. Cardiac output is an additive control in the model. Cardiac output increases to maintain arterial pressure x_1 with decreasing disturbances in systemic resistance due, for example, to hypoxia, anemia, exercise, or heat. Also, resistance elements are adjusted to maintain arterial pressure x_1 with low cardiac output from such disturbances as gravity, stress, shock, and hemorrhage.

2.3.2 Compartmental Processes

Similar compartmental models have been utilized successfully to study such diverse processes as thermal regulation (Section 2.2, Vol. 1), body fluid and electrolyte balance, kinetics of metabolites in cell suspensions or tissues, circulating thyroxin [D3], dynamics of material ingested to and excreted from the body, control of carbon dioxide in the body (Section 2.3.1), and the distribution of lymphocytes to various organs and lymph nodes of the immune system [M14].

Consider two compartments of capacities c_1 and c_2, uniform concentrations $\xi_1(t)$ and $\xi_2(t)$, and amounts $x_1(t)$ and $x_2(t)$ of substance X. Then the net flux diffusing through the membrane from compartment 2 to compartment 1 is described by

$$\phi_{12} = \rho_{12}\xi_1(t) - \rho_{21}\xi_2(t), \tag{2.12}$$

or

$$\phi_{12} = \frac{\rho_{12}}{c_1}x_1(t) - \frac{\rho_{21}}{c_2}x_2(t) \tag{2.13}$$

and

$$\phi_{12} = -\phi_{21}, \tag{2.14}$$

where ρ_{21} and ρ_{21} are membrane exchange coefficients (not generally of the same value). It is assumed that the membrane has no storage of substances [M2].

In addition to the natural diffusion mechanism for material transport from regions of high concentration to regions of low concentration, experimental evidence shows that sometimes there exists an active transport from low concentration to high concentration or a retention of high concentration of various substances by cells against appreciable gradients. The necessary energy for this active transport is received from the cellular metabolism. The net active transport of a single substance across a membrane may be modeled by

$$\frac{a(t)\xi_2(t) - b(t)\xi_1(t)}{c + d\xi_1(t) + e\xi_2(t)}, \tag{2.15}$$

where it is assumed that a and b may be manipulated, and c, d, and e are constants. While active transport is obviously nonbilinear it may be approximated by the linear terms of a Taylor series so that the overall process is approximated by a system of bilinear equations with $a(t)$ and $b(t)$ as control variables and $\xi_1(t)$ and $\xi_2(t)$ as state variables. Similarly, material transport due to hydrostatic pressure differences may be closely approximated by a bilinear model.

Suppose that a single substance X is distributed among n compartments (cells, spaces, phases, pools). The rate of change of substance amount in the ith compartment is described by the conservation equations:

$$\frac{dx_i}{dt} = \sum_{\ell=1}^{n}{}' \phi_{i\ell}(t) - \sum_{q=1}^{n}{}' \phi_{qi}(t) + \phi_{ia}(t) - \phi_{ai}(t) + p_i(t) - d_i(t),$$
$$i = 1, \ldots, n, \tag{2.16}$$

where the primed summation sign indicates that the ith term of the summation is deleted, $x_i(t)$ is the amount of substance X in compartment i, ϕ_{ij} the flux of X from compartment j to compartment i, $\phi_{ia}(t)$ [$\phi_{ai}(t)$] the flux of X into [out of] compartment i from [to] the environment, and $p_i(t)$ [$d_i(t)$] the rate of production [destruction] of X in compartment i.

Equation (2.16) may be put into a more compact form so that environmental fluxes, production terms, and destruction terms are combined with compartmental fluxes. In this manner,

$$\frac{dx_i}{dt} = \sum_{\ell=1}^{n+2}{}' \phi_{i\ell}(t) - \sum_{q=1}^{n+2}{}' \phi_{qi}(t), \qquad i = 1, \ldots, n, \tag{2.17}$$

where $\phi_{i,n+1}(t) = \phi_{ia}(t)$, $\phi_{n+1,i}(t) = \phi_{ai}(t)$, $\phi_{i,n+2}(t) = p_i(t)$, and $\phi_{n+2,i}(t) = d_i(t)$.

If it is assumed that compartmental capacities do not change significantly over the time period of interest [see (2.13)], the flux terms can be expressed in terms of time-invariant bilinear equations of the general form of

$$\phi_{i\ell}(t) = \sum_{j=1}^{n} a_{i\ell,j}x_j(t) + \sum_{j=1}^{n}\sum_{k=1}^{m} n_{i\ell,jk}x_j(t)u_k(t) + \sum_{k=1}^{m} b_{i\ell,k}u_k(t) + v_{i\ell}(t) \tag{2.18}$$

for $i, \ell = 1, \ldots, n+2$, $i \neq \ell$; $i, \ell \neq (n+1, n+2)$ or $(n+2, n+1)$. For some problems $v_{i\ell}(t)$ may be neglected or may be a given time function. In other cases, production and destruction terms may be additive controls. To distinguish this sytem as a slightly different class of bilinear system, let the affine bilinear model be described by

$$\frac{d\mathbf{x}}{dt} = \mathbf{Ax} + \sum_{k=1}^{m} \mathbf{N}_k u_k \mathbf{x} + \mathbf{Bu} + \mathbf{v}, \tag{2.19}$$

where $\mathbf{v}$ is an n-vector and the other terms are as described previously. Equations (2.17) and (2.18) are put into this general bilinear form if $\mathbf{A}$ has elements

$$a_{ij} = \sum_{\ell=1}^{n+2}{}' a_{i\ell,j} - \sum_{q=1}^{n+2}{}' a_{qi,j},$$

$$n_{i,jk} = \sum_{\ell=1}^{n+2}{}' n_{i\ell,jk} - \sum_{q=1}^{n+2}{}' n_{qi,jk}, \qquad k = 1, \ldots, m,$$

$$b_{ik} = \sum_{\ell=1}^{n+2}{}' b_{i\ell,k} - \sum_{q=1}^{n+2}{}' b_{qi,k},$$

$$v_i = \sum_{\ell=1}^{n+2}{}' v_{i\ell}(t) - \sum_{q=1}^{n+2}{}' v_{qi}(t).$$

Here Σ' indicates the constrained summations $\ell, q \neq i$. While real processes may involve transport of more than one substance, no loss of generality is obtained in regard to the bilinear form of the system equations above.

The concept of compartmentation has been applied for anatomical and mathematical convenience here. Still this does not imply that optimal discretization of the distributed system in a mathematical sense would yield the same compartments. In the analysis above, it is implicitly assumed that the process involves mass transport. Again the concept is more general. By their nature, these systems are difficult to isolate for purposes of direct identification and modeling. It is frequently impossible to observe state or control variables, yet manipulate control variables as is so often desired for system identification. The compartmented structure of these systems, however, does lend itself conveniently to tracer analysis, which in turn can be used for system identification.

A tracer substance in effect puts a tag on some quantity for which it is desired to study dynamic behavior. Isotope tracers are particularly convenient and usually harmless to identify compartment capacities and substance amounts and concentrations. Usually, it is assumed (1) that the tracer is distributed uniformly throughout the substance X, (2) that labeled and unlabeled X behave identically, and (3) that the tracer does not alter the system dynamics. Obviously, this requires that the amount of tracer in the system be neglected. The portion of tracer in the total amount of X in compartment i, called the specific activity a_i, is described by

$$\frac{d(x_i a_i)}{dt} = \sum_{\ell=1}^{n}{}' \phi_{i\ell}(t)a_1 - \sum_{q=1}^{n}{}' \phi_{qi}(t)a_i + \phi_{ia}(t)a_{ia}(t) - \phi_{ai}(t)a_i - d_i(t)a_i + f_i(t),$$
$$i = 1, \ldots, n, \qquad (2.20)$$

where $a_{ia}(t)$ is the specific activity of the substance in influx $\phi_{ia}(t)$, and $f_i(t)$ is the influx of tracer inserted directly into compartment i. If $\phi_{ia}(t) \equiv 0$, then $a_{ia}(t) \equiv 0$. It is interesting and convenient that the tracer system (2.20) is linear in specific activity even if the physiological model is nonlinear. Of course, this linear structure also constrains its application in system identification.

Ordinarily, tracer experiments are conducted when the compartmental system is in steady state with $\dot{x}_i = \dot{\phi}_{ij} = \dot{\phi}_{ai} = \dot{\phi}_{ia} = \dot{p}_i = \dot{d}_i = 0$, $i, j = 1, \ldots, n$. Therefore, the unforced tracer dynamics are time-invariant, linear, and take the form

$$\frac{d\mathbf{a}}{dt} = \mathbf{Sa} \qquad (2.21)$$

where $\mathbf{a}$ is a column vector of components $a_1, \ldots, a_n$, and $\mathbf{S}$ is an $n \times n$ matrix of elements

$$S_{ij} = \frac{\phi_{ii}}{x_i} \geq 0, \qquad i, j = 1, \ldots, n, \quad i \neq j,$$
$$S_{i\ell} = \frac{1}{x_i}\sum_{\ell=1}^{n}{}' \phi_{i\ell} = -\frac{1}{x_i}\sum_{q=1}^{n}{}' \phi_{qi}, \qquad i = 1, \ldots, n. \qquad (2.22)$$

For most physiological processes it is not at all convenient to decouple subsystems, to measure all controls, and to measure all state variables. Usually, it is not practical to add test signals to the physiological process that is to be modeled. Unfortunately, this precludes the use of many of the most convenient properties of bilinear systems. Such limitations are especially prevalent in human-body experiments. For many of these processes, compartmentation lends itself conveniently to modeling and identification by means of tracer experiments since such experiments may be conducted without having a noticeable effect on the regulated process. Added to this is the convenience of linearity exhibited by the tracer kinetics above. Tracer dynamics is investigated further in Section 3.6 of Vol. 1 for compartmental system application. Tracer can be inserted by special routes, such as by injection, directly into a particular compartment, or tracer can be ingested by normally existing pathways. Here, suppose that tracer is inserted directly into q system compartments and that its behavior can be monitored in p compartments. Define an input matrix $\mathbf{Q}$ with unity elements in each row corresponding to a compartment accessible for tracer insertion and with zero elements elsewhere. Each column of $\mathbf{Q}$ has exactly one unity element and the rank of $\mathbf{Q}$ is q. Also, define a $p \times n$ output matrix $\mathbf{P}$ with zero elements except for unity in each column, which corresponds to an observable compartment. Then from (2.21) or (2.22) experimental access is described by

$$\frac{d\mathbf{a}}{dt} = \mathbf{Sa} + \mathbf{X}_d^{-1}\mathbf{Qf}_i$$

and

$$\mathbf{w} = \mathbf{Pa},$$

where $\mathbf{X}_d$ is the diagonal matrix of elements $x_1, \ldots, x_n$ (assumed to be at steady state); $\mathbf{f}_i$ is the q-dimensional vector of inserted total activity fluxes, and therefore $\mathbf{X}_d^{-1}\mathbf{Qf}_i$ is the vector of inserted specific activity fluxes; and $\mathbf{w}$ is the p-dimensional output vector of observed compartmental specific activities.

2.4 THE IMMUNE SYSTEM

Engineers embroiled in large-scale network problems may well appreciate immunological complexes whereby the humoral subsystem alone includes some 10^{12} lymphocytes and 10^{20} antibody molecules of 10^{7} different chemical affinities in the immunologically unstimulated human body. Unfortunately, linear system theory, so significant in electrical networks, is shown to have limited application to immunology. On the other hand, coupled bilinear control systems arise due to the nature of cell division, differentiation, and chemical binding. Similarly, a chemical cascade of nonlinear amplifiers is generated as the complement subsystem for such purposes as tumor cell lysis. This controlled use of instability by means of parametric manipulation, which is so effective in disease control, may very well provide a clue for new design methodologies in engineering practice.

It is anticipated that system methodology will help bring some focus to the overall program of immunological research. Indeed, the results that can be expected from the application of the system approach to immunology are numerous, but the following come to mind immediately [M14]:

1. Modeling and identification theory should provide a deeper and more quantitative description of the immune process and the related experimental procedures; it should also help the critical analysis of hypotheses, the understanding of biological mechanisms, the prediction of behaviors, and the design of experiments.
2. The analysis of the structural properties of the models, such as stability, observability, and controllability, which are peculiar of system theory, should characterize some meaningful physical properties of the process itself.
3. Control and optimization theory should stimulate new and more satisfactory approaches to the vaccination policies, the treatment of diseases that are under immunological influence (such as infections and tumors), and the control of reject phenomena in organ transplantation and prosthesis.
4. Filtering, estimation theory, and data processing methods should allow a more sophisticated and complete recovery of information from experimental results.

Of course, the actual availability and development of adequate automatic computers and the related software is a key point in pursuing the previous objectives. Still, the methodologies of (1) have already found a large extent of application; also, some results are available on (2) and (4). Applications in the line of (3), however, are very scarce at the present time.

Only a brief description of the immune system can be presented here, but several excellent texts are available (e.g., [R4] and [E3]). The body's immune system has the task of controlling alien substances. These may take the form of bacteria, viruses, macromolecular chemical contaminants, and toxins. Then, by complex recognition mechanisms, which discriminate between "self" and "nonself" uniquely on the basis of the chemical structure, the immune process generates cells and molecules to bind and to destroy the alien. These defenders are circulated throughout the various body organs by the bloodstream, through tissue, appropriately processed in their migration and recirculated back through the lymphatic vessels. The generation of these defender cells and molecules is termed the *immune response*, while any alien substance that is able to induce such a response is called the *antigen* (Ag).

Although the host's defense mechanisms include many interrelated processes, the generation of molecules called *antibodies* (Ab) from various cellular interactions (*humoral* immune response) will be emphasized here. The binding of antibodies to an antigen facilitates its removal, which is carried out by specialized cells of the organism (phagocytes or macrophages). In the case of antigens in molecular form, in fact, the complexes of antigen and antibodies (called immune complexes) may form aggregates of large dimensions, endowed with cytophilic properties. Further, the defense function of antibodies may happen by direct neutralization of the antigen activity (chemical or biological). When the antigen is a bacterium or a virus contaminant, the binding of antibodies may lead to the lysis of the invader. After the antibody binding to the bacterium surface, a cascade of protein enzyme reactions may be triggered, involving the nine components of certain serum proteins called complement, which can lead to the rupture of the bacterial wall. Functionally, the process is a sequence of multiplicative control amplifications of enzyme concentration to drill through the cell wall.

While cytotoxic T cells seem to engulf and destroy (phagocytosis) certain antigens directly, lysis by complement and opsonization by macrophages do require the antibody tag. The overall collaboration in defense between phagocytes, antibodies, and complement is shown in Fig. 2.1. In these experimental results, the death rate of bacteria is increased by nearly an order of magnitude by first the addition of sufficient antibody and then again by the addition of sufficient complement. Consequently, the time constant, as depicted by the line slope in Fig. 2.1, is set by a multiplicative control action in the model.

Antibodies [*or immunoglobulins* (Igs)] are molecules that, by means of *binding sites*, may bind to certain chemically defined small portions of the antigen, called *antigenic determinants* (or epitopes). The equilibrium association constant of the binding with a certain antigenic determinant is termed *affinity* of the antibody molecule for the related antigen. Approximately 20 percent of plasma protein is in the form of this heterogeneous group.

The cells of the lymphatic system (see Fig. 2.2) circulate in the body and are present in high concentration in certain lymphoid organs, such as spleen and lymph nodes. Two lineages of lymphocytes may be recognized: B lymphocytes and T lymphocytes. B cells are directly bone marrow derived, while T cells attain their maturity passing through the thymus. In the humoral immune response, B lymphocytes, upon *stimulation* by antigen, undergo a process of *proliferation* and *differentiation* into long-living lymphocytes called *memory cells*, and short-lived but high-antibody-secreting cells, called *plasma cells*. Antibodies are secreted also, but to a lesser extent, by cells incompletely differentiated. The function of T cells in the humoral response is the control of B cells and their production of antibodies. For this function there is experimental evidence as to the existence of T-suppressor as well as T-helper (stimulatory) cells.

Macrophages (M_ϕ), which are also generated by bone-marrow stem cells, beyond their role in removing antigen directly, collaborate with B and T cells in stimu-

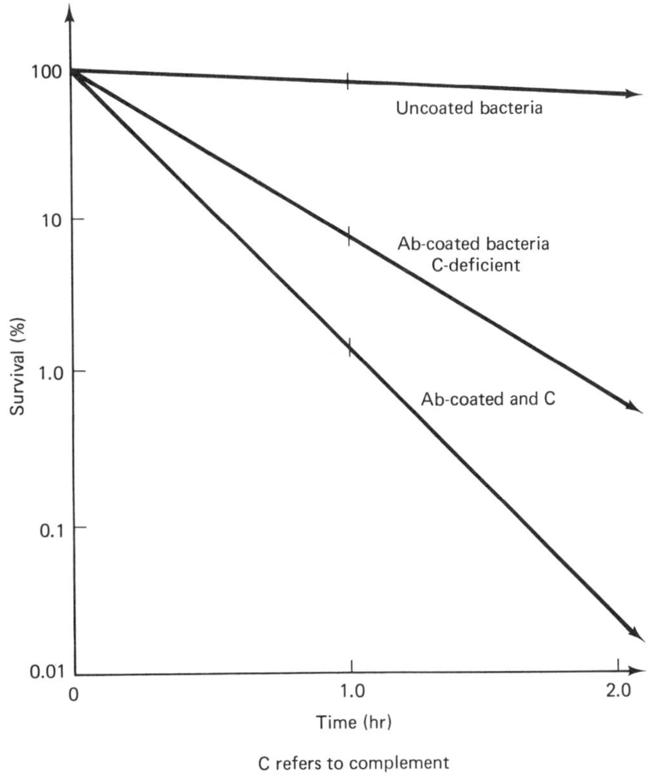

Figure 2.1 Bacteria control in serum.

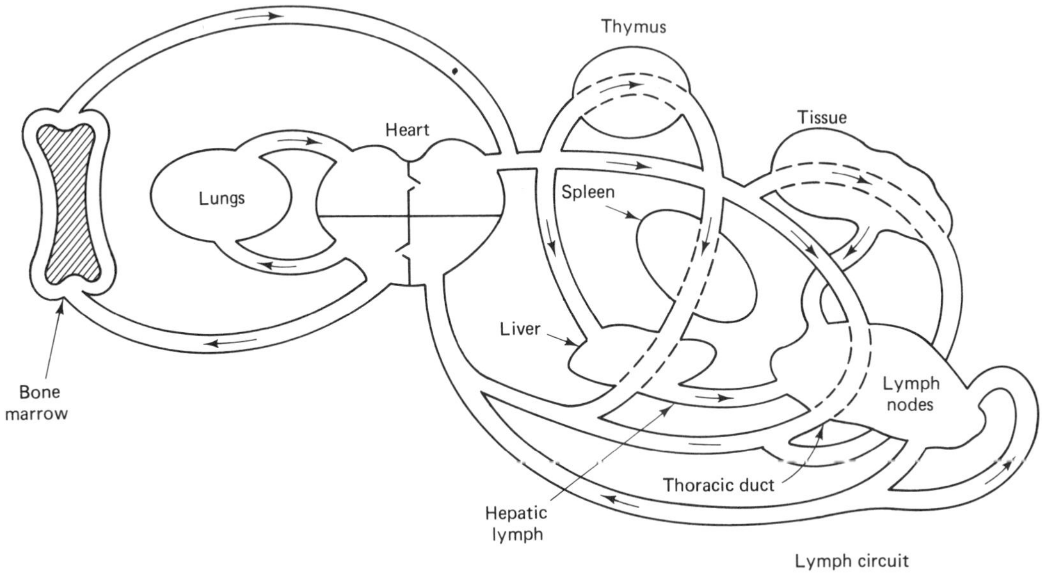

Figure 2.2 Lymphocyte circulation component of the immune system.

cells.) The migration coefficients, $\delta_{i,s,u}(\cdot)$ and $\delta_{i,r,s}(\cdot)$ in general could be deterministic or stochastic functions of appropriate $y_m(t)$ since certain lymphokines (e.g., macrophage migration inhibition factor, MIF lymphokine) manipulate migration coefficients. A similar molecular version of (2.26) may be developed. Note the similarity of the last two terms in (2.26) to the right-hand side of (2.12).

Similar to the cellular model (2.23), a conservation equation would yield the following for *tumor-cell population* $z(t)$:

$$\frac{dz}{dt} = v_z(t) + p_z(\cdot)z - \frac{z}{\tau_z} + \sum_j 2p_j(\cdot)p_{jz}(\cdot)x_j - u_z(\cdot)z, \tag{2.27}$$

where $v_z(t)$ is an external source of malignant cells (such as from a metastasized tumor); $p_z(\cdot)$ and $p_j(\cdot)$ are appropriate growth coefficients (including probability of stimulation for division); and $p_{jz}(\cdot)$ is a differentiation coefficient (probability). All three coefficients may be functions of z itself, certain lymphokines, and interferon (IFN). However, they commonly may be assumed constant. $p_{jz}(\cdot)$ considers the possibility of a normal lymphocyte becoming malignant. $u_z(\cdot)$, the parametric control, is a function of concentrations (populations) of T_c, M_a, and possibly Ab, N_k (natural killers), C (complement), IFN, and z itself. For example, $u_z(\cdot) = k_z(x_M + x_T)/(z + L_z)$ would seem appropriate if any control-limiting constraint is handled by the denominator term with k_z and L_z appropriate constants, and $x_M(t)$, $x_T(t)$ concentrations of angry macrophages (M_a) and cytotoxic T cells (T_c), respectively, assuming that these are the major killers in question. τ_z is the natural death time constant of the tumor. Migration of malignant cells from the tumor could be included in the net source term $v_z(t)$ or similar to above compartmental model.

Tumor debris of concentration x_d might be generated to enhance tumor recognition by the immune mechanisms. This would be modeled by the latter term in (2.27) with a natural death or removal time constant τ_d so that

$$\frac{dx_d}{dt} = u_z(\cdot)z - \frac{x_d}{\tau_d}. \tag{2.28}$$

Consequently, the tumor antigen concentration $\rho_a(\cdot)$, as presented for stimulation of helper T cells (T_h) by the macrophages (M_ϕ and M_a) of concentrations x_ϕ and x_M, might be approximated by

$$\rho_a = \alpha_d(\cdot)(x_\phi + x_M), \tag{2.29}$$

which according to (2.23) would be proportional to a p_{sh} coefficient of differentiation from stem to T_h cells. Here α_d might be limited according to

$$\alpha_d = \frac{c_d x_d}{L_d + x_c}$$

with c_d and L_d appropriate constants. Again stochastic parameters may be utilized depending on the accuracy desired and the information available (through theory and data). More commonly, $\alpha_d \approx c_d x_d$. The structure of x_ϕ and x_M dynamics follows (2.23).

While the manipulated parameters in this tumor model assume certain cellular feedback, the actual feedback, for the most part, involves appropriate molecular factors which are assumed proportional to the cell concentrations secreting them.

Several complications to tumor control need to be studied more thoroughly to be modeled accurately through parametric "control" terms as above but with added equations in some cases. These consist of tumor blocking, tumor escape (metastases), relevant lymphokines, and conversion of healthy cells to tumor cells. In general it is seen that the *immune system structure* consists of numerous multiplicatively (parametrically) coupled systems. Or, they are structured from basic bilinear system (BLS) components, which, in some cases, are coupled in a nonlinear fashion to model limiting. Fortunately, considerable theory has been derived for the analysis of BLS. Some of this may be useful in terms of understanding individual BLS immune elements, such as the multiplicative effects of controlling bacteria. Indeed, the drastic controllability requirement of the immune system suggests bilinear control of stable and unstable modes as borne out by the conditions for completely controllable BLS in Chapter 4. If the appropriate outputs are measured (i.e., those which formulate the multiplicative controls or inputs), the coupled BLS above become conditionally linear, and thus called conditionally linear systems CLS. Even with nonlinear parametric coupling, the CLS structure is retained if those states formulating the nonlinear parameters are measured. *

2.4.2 The Humoral Process

Consider the cellular molecular structure above for the case of the generation of antibodies in response to antigen stimulation (i.e., the *humoral immune system*). Here only one class of antibodies is considered and memory cells are lumped with immunocompetent cells. Still the model does describe the general behavior of the humoral system in at least a crude fashion. Rather than use $y_m(t)$ for the molecular concentrations, $x_i(t)$, $i = 1, \ldots, 5$, is used to represent all state variables.

The process is described by the following variables, with association-constant (k) dependence omitted for brevity or for a single k approximation.

$x_1(t)$ Population concentration of immunocompetent cells (ICC), which are sensitized lymphocyte cells with particular surface receptors having association constant of affinity k for antigen. They may differentiate into plasma cells and divide into memory cells. The latter, which may further divide, enter the pool of immunocompetent cells (ICC pool includes all memory cells in this model).

$x_2(t)$ Population concentration of short-lived plasma cells which are nonreproducing offspring of stimulated immunocompetent cells.

$x_3(t)$ Population concentration of free "antibody sites" (unbound to antigen). Antibodies are considered having independently acting sites.

$x_4(t)$ Population concentration of immune complexes, which individually include antibody sites and antigen, bound together according to mass-action chemistry.

$x_5(t)$ Antigen concentration, unbound to antibody, which triggers the response mechanism after sufficient binding and cellular interactions. Antigen is considered as functionally univalent for convenience.

Then, from (2.23) and (2.24), the humoral system equations may be derived directly for the cells and molecules of concern. To better understand the cell process, however, first consider the ICC dynamics from basic population continuity. Immunocompetent cells first lead to a change of population density Δx_1 on an interval Δt due to mitosis, given by

$$[2\alpha(1 - p_s)p_d x_1 - \alpha p_s x_1]\,\Delta t, \tag{2.30}$$

where α is the birthrate constant of stimulated immunocompetent cells, p_d the probability that an immunocompetent cell differentiates into a plasma cell (x_5, k dependent), p_s the probability that antigen stimulates a cell (x_5, k dependent), and the argument t is dropped from the state variable for brevity.

To this must be added the source of new stem cells $v_1\,\Delta t$, and the death of cells $x_1\,\Delta t/\tau_1$, where v_1 is the rate of generation of new immunocompetent cells (from bone marrow), and τ_1 is the mean lifetime (or time constant) of immunocompetent cells. While v_1 may be a function of t for prescribed immunotherapy some day, it is normally assumed constant though actually distributed over k. In the limit, the process is described by an ordinary differential equation of form

$$\frac{dx_1}{dt} = \alpha u_1 x_1 - \frac{x_1}{\tau_1} + v_1, \tag{2.31}$$

where $u_1 = p_s(1 - 2p_d)$. To be consistent with (2.23), $p_1 = \alpha p_s$, $p_{12} = p_d$. Also, x_1 loss due to x_2 transformations is neglected. Similarly, plasma cells, antibody sites, and immune complex generation for each k, may be approximated by birthrate minus death rate (including loss to mass-action binding) equations as follows:

$$\frac{dx_2}{dt} = 2\alpha u_2 x_1 - \frac{x_2}{\tau_2}, \tag{2.32}$$

$$\frac{dx_3}{dt} = -cu_3 x_3 - \frac{x_3}{\tau_3} + \beta_{23}x_2 + cx_4 + \beta_{13}x_1, \tag{2.33}$$

$$\frac{dx_4}{dt} = cu_3x_3 - \left(c + \frac{1}{\tau_4}\right)x_4, \tag{2.34}$$

where τ_2, τ_3, and τ_4 are approximate lifetimes (or time constants); $u_3 = kx_5$ and $u_2 = p_sp_d$; β_{23} is the plasma-cell antibody production rate; β_{13} the ICC antibody production rate; and c and kc the dissociation rate and association rate coefficients, respectively, of the immune complex (k-dependent). The state equations above represent total concentrations at a given average affinity k or represent only that concentration corresponding to one of many k. In the latter case, separate equations could be written for each individual k or for several significant values.

The free antigen population density appears as an input to which the immune control system responds. The antibodies secreted after inoculation of the antigen population at a rate $v_2(t)$ include a distribution over association constants which would require a summation over the appropriate values of k in the model. The rate of generation of a free antigen concentration may be approximated by the inoculation rate minus rate of catabolism minus net complex association rate.

$$\frac{dx_5}{dt} = v_2 - \frac{x_5}{\tau_5} - Nc(u_3x_3 - x_4), \tag{2.35}$$

where τ_5 is the average antigen lifetime. N is a weighting constant to account for the total number of k which actually replaces a summation of terms. In the notation of (2.23), $c = c_{34} = c_{43} = c_{45}$, $u_3 = x_5$, $Nc = c_{54}$.

It is of interest that the B-cell model has the following control structure:

$$\frac{d\mathbf{x}}{dt} = \mathbf{A}\mathbf{x} + (\mathbf{b}_1u_1 + \mathbf{b}_2u_2)x_1 + \mathbf{b}_3u_3x_3 + \mathbf{c}_4v_1 + \mathbf{c}_5v_2, \tag{2.36}$$

where state vector $\mathbf{x} \in \mathbb{R}^5$, and multiplicative feedback controls $\mathbf{u} \in \mathbb{R}^3$,

$$\mathbf{A} = \begin{bmatrix} -\frac{1}{\tau_1} & 0 & 0 & 0 & 0 \\ 0 & -\frac{1}{\tau_2} & 0 & 0 & 0 \\ \beta_{13} & \beta_{23} & -\frac{1}{\tau_3} & c & 0 \\ 0 & 0 & 0 & -\left(c + \frac{1}{\tau_4}\right) & 0 \\ 0 & 0 & 0 & Nc & -\frac{1}{\tau_5} \end{bmatrix},$$

$$b_1 = \begin{bmatrix} \alpha \\ 0 \\ \vdots \\ 0 \end{bmatrix}, \quad b_2 = \begin{bmatrix} 0 \\ 2\alpha \\ 0 \\ 0 \\ 0 \end{bmatrix}, \quad b_3 = \begin{bmatrix} 0 \\ 0 \\ -c \\ c \\ -Nc \end{bmatrix}, \quad c_4 = \begin{bmatrix} 1 \\ 0 \\ \vdots \\ 0 \end{bmatrix}, \quad c_5 = \begin{bmatrix} 0 \\ \vdots \\ 0 \\ 1 \end{bmatrix}.$$

While the system consists of several coupled BLS, the closed-loop immune system is more highly nonlinear but is conditionally linear for appropriate measurements. The *additive control* v_1 is independent of u and is significant in *immunotherapy*. The model can be used to further analyze the effectiveness of such treatment by consideration of deterioration of the infected process or organ as an added component to the model. Though this source of stem cells, v_1, is distributed according to affinity k (usually assumed to be Poisson or Gaussian), an average k seems representative. Here the probability that antigen stimulates the cells of affinity k depends on the number of appropriate sites that are occupied. A simulation of system (2.36) for an antigen in a complete Freund adjuvant (oil base) injected into mice has been compared quite favorably with experimental data [M15].

Although the closed-loop humoral immune model is highly nonlinear, it is felt that the open-loop "plant" and its structure are also significant with respect to a systematic understanding and to future research on immunotherapy or disease control. It may be convenient to turn on (or off) parts and/or mechanisms, which are most effective against (or protective of) a particular alien, by drug manipulation of p_s, p_d, or other multiplicative feedback "controls." Indeed, the immune system can be thought of as various subsystems (often coupled BLS) which are turned on and off by numerous mechanisms such as T-helper and T-suppressor cells. In a sense, the process makes controlled use of instability to enhance controllability, which is so popular in BLS. It will be shown in Chapter 4 that such parametric control can improve controllability and system performance over that of additive control alone. Furthermore, it should be noted that these feedback mechanisms are the least understood. While u_3, binding multiplier is probably the best understood of the three, it may be modeled by complex functions to show a saturation effect. But with respect to the nature of u_1, u_2, (or p_s, p_d), there is considerable controversy.

2.4.3 Disease

Consider the deterioration of organ tissue from a virus of concentration V by [M16]

$$\frac{dm}{dt} = \sigma V - H_m m, \tag{2.37}$$

where m is a relative damage characteristic of the tissue, σ a virus damage coefficient, and H_m a tissue recovery coefficient. In this model, which follows [M16], V is described similar to x_5 in (2.35) but with an added virus (antigen) destruction term by cytotoxic T cells. Model state variables include equivalent concentrations of virus (antigen), antibody, B-ICC, plasma B cells, and m (above). The Marchuk model takes the form of the above, without x_4, and with $\alpha = 0$ so that

$$\frac{dx_3}{dt} = \alpha'\xi(m)\, x_2 - ckx_3x_5 - \frac{1}{\tau_3}x_3, \tag{2.38}$$

where $\xi(m)$ is an appropriate monotone decreasing function. An analysis of the time response for this model is presented in [M16]. Similar equations were developed for various T cells and macrophages, which in turn affect feedback controls.

Further investigations show the relation of stationary solutions to healthy organ state and to chronic disease. The latter may also result in periodic solutions [M16]. It is shown by simulation that sufficiently increasing the initial virus dose results in more effective control of the chronic disease by the immune system—a treatment that has had successful clinical application to one of the modelers (Academician Marchuk).

Acquired immune deficiency syndrome (AIDS) has been targeted as the most life-threatening disease of the future. During the 1980s the number of cases seems to be increasing almost as an uncontrolled exponential function. The intent here is to describe the progressive nature of the disease and the resulting loss of immunocompetence. Hopefully, such a model may be useful to control AIDS (i.e., to derive a treatment to minimize its effect). It has been shown that AIDS is linked with a virus called HIV, human immunodeficiency virus. HIV has a particular affinity for a certain class of T-cell lymphocytes, T_4, which probably may be lumped with T-helper cells. It is reasonable to expect that the eventual death of the T_4 lymphocytes from the viral infection may be a principal inducement of immunocompetence. While the HIV does infect macrophages, experimental evidence indicates that their death rate from this virus is not drastic enough to cause a deterimental loss of immunocompetence.

AIDS dynamics may be approximated by (2.23) and (2.24) with a few example equations provided here. First, the antibody generation is approximated by the humoral model (2.31)–(2.36) with a more complex $u_1(\cdot) = u_1(x_5, x_T, x_\phi)$, $u_2 = u_2(x_5, x_T, x_\phi)$, where x_T and x_ϕ refer to concentrations of free T-helper cells, T_h, and free macrophages, M_ϕ. x_5, the antigen concentration here refers to the HIV concentration.

Although appropriate molecular interactions are responsible for HIV control it is assumed proportional to the primary molecular source x_T for simplicity. Any HIV control effects from x_ϕ, x_3 are neglected, despite their increases, which is consistent with present experimental data. Such data suggest that T_h-generated $CD4$ molecules do have some detrimental effect on HIV.

If the humoral antibody and macrophage components are neglected, and HIV is represented through its infected T-cell component (with the infected macrophages neglected), the dynamical structure may be approximated as follows:

$$\dot{x}_5 = v_3 + \left(\alpha_5 - \frac{1}{\tau_5}\right)x_5 + \beta_{T5}x_T' - c_{T5}'x_5x_T - \mu_4x_5, \tag{2.39}$$

where again v_3 is the source rate of HIV infection from the environment; the second term is a net birthrate term, $\beta_{T5}x_T'$ is the generation rate from infected T_h cells of concentration x_T'; the last two terms are the rate of HIV infectious binding with free

T_H cells and T_H-cell control term, respectively. The latter may be approximated by $u = c_{T5}x_T$, where c_{T5} is an effective positive association constant (via $CD4$ molecules).

Consequently, the free T_h-cell dynamics may be approximated by

$$\dot{x}_T = u_5 x_T - \frac{x_T}{\tau_T} - u_4 x_5 + v_4, \tag{2.40}$$

where u_5 is a stimulation coefficient function of antigen concentration and possibly other molecular controls. v_4 is the source rate of free helper cells, which in reality would involve several stages of precursors. Similarly, the infected-T_h dynamics become

$$\dot{x}'_T = u_6 x'_T - \frac{x'_T}{\tau'_T} + c'_{T5} x_5 x_T + v_5, \tag{2.41}$$

where u_6, v_5, τ'_T, c'_{T5} are appropriately defined similar to the corresponding term in (2.39).

A study of a simpler AIDS model, but which assumes $x_T(t)$ to be a random process, is studied by Merrill [M17]. A simple deterministic AIDS model is also studied by Cooper [C4]. The two models are compared in [M17] with the deterministic model projecting the most rapid deterioration of immune response.

References [B8], [H9], [M14], [M16], and [P2] provide a good summary of recent research on system immunology. The increasing popularity of this topic is exemplified by the rapidly increasing number of publications, conference sessions, and workshops. In 1987 alone such workshops were sponsored in Santa Fe by the U.S. Department of Energy and in Vienna by the International Institute for Applied Systems Analysis (IIASA). Numerous previous meetings or sessions were sponsored by IIASA, the International Federation of Information Processing (IFIP), the Institute of Electrical and Electronic Engineers (IEEE), the U.S.S.R. Academy of Sciences, the U.S. National Science Foundation, and the Italian Council of National Research. The most significant advances in applying such models to hospital patient care has taken place in Moscow under the direction of the U.S.S.R. Academy of Sciences President Marchuk. Parameter identification for the humoral process by means of Walsh functions is studied in Section 5.4.2.

2.5 MISCELLANEOUS PROCESSES AND COMMENTS

Only a small representative sample of physical systems and their mathematical models have been studied here. A complete treatment of this topic is beyond the scope of this book. Still it is hoped that the reader, at this point, will have a better appreciation for this most significant and most complicated problem. And while linear systems are convenient for analysis, they are not all that common in the real world. Only for academic problems in general can one expect to have plant transfer functions or plant state and control coefficient matrices provided. Chemical process control, electric power networks, ecology, and socioeconomics are most significant areas which have not been studied here due to space limitations. However, the same fundamental concepts of conservation, the Euler–Lagrange equation, and others are still useful. Some of these processes will be introduced along with examples throughout the book. Compartmental models can be useful in chemical process control and socioeconomic regional modeling as well as in biological systems which were noted above. The same structure may arise with tracer experiments, including appropriate polls as well as chemical or isotopic tracers.

EXERCISES

2.1 Show that if impedance is a control variable, a simple, passive, electrical network becomes a BLS.

2.2 Derive a BLS perturbation model for $\delta\mathbf{x}(t)$, $\delta\mathbf{u}(t)$ at an equilibrium state-control reference $(\mathbf{x}^*, \mathbf{u}^*)$. Here $\dot{\mathbf{x}} = \mathbf{f}(\mathbf{x}, \mathbf{u})$, $\delta\mathbf{x}(t) = \mathbf{x}(t) - \mathbf{x}^*$, and $\delta\mathbf{u}(t) = \mathbf{u}(t) - \mathbf{u}^*$.

2.3 Determine the sufficiency of the compartmental flow, tracer insertion, and tracer measurement configurations shown in Fig. 2.3 from a linear system minimum realization test (controllability and observability).

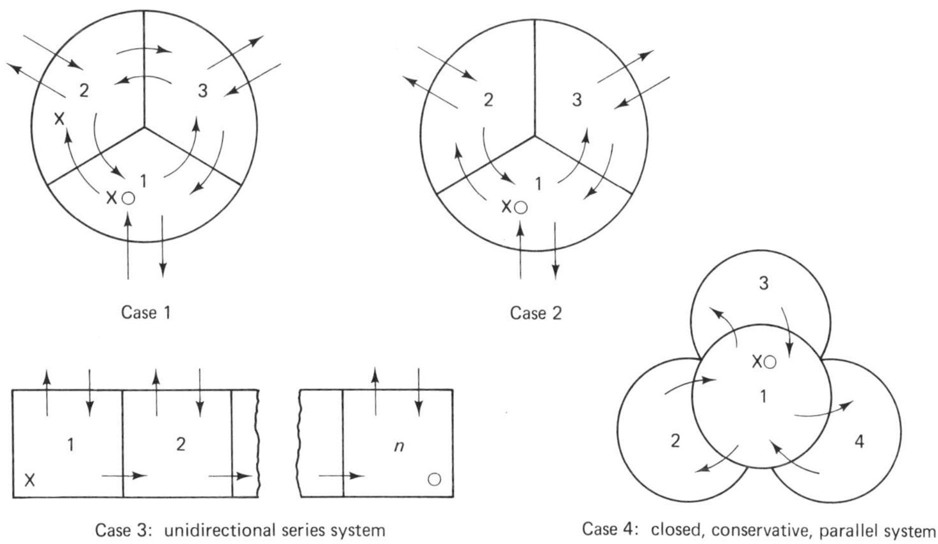

Figure 2.3 Compartmental tracer-flow configurations for Exercise 2.3. Arrows denote flux; x, tracer insertion; o, tracer observation.

2.4 Derive a tumor-growth control model from (2.23) and (2.27)–(2.29) with

x_1-stem-cell population	x_5-M_a population (x_a)
x_2-T_h-cell population	x_6-tumor debris level (x_d)
x_3-T_c-cell population (x_T)	x_7-tumor cell population (z)
x_4-M_ϕ population (x_ϕ)	

Use the cell conservation principle and assume nonzero p_1, p_2, p_4, p_7, p_{12}, p_{13}, p_{14}, p_{45}, c_{37}, c_{73}, c_{57}, c_{75}, v_1, τ_i, $i = 1, \ldots, 7$. p_i, p_{ij} are appropriate experimentally determined nonlinear functions of x_k, $k = 2, \ldots, 7$, which depend on interleukin factors, interferon, and inflammation. Other parameter values are assumed to be zero. c_{ij} refer to direct cellular binding constants for association or dissociation since further molecular details are neglected. Tumor debris, x_6, is generated at a rate $\alpha_{67}x_7$—its natural death rate + the binding rate of tumor cells with T_c and M_a.

2.5 The circulation of lymphocytes in the immune system generally is rapid compared to their stimulation, consequent differentiation, and death, which leads to an immune response. With such an assumption derive a simple dynamic model for the organ distribution of recirculating lymphocytes shown in Fig. 2.2.

CHAPTER 3

BILINEAR SYSTEM STABILITY

3.1 INTRODUCTION

The second or direct method of Liapunov proved useful for the stability analysis of nonlinear systems in Vol. 1, Chapters 6 and 7. In the present chapter, the method is applied to bilinear systems, BLS—an important class of nonlinear systems which has some of the features of linear systems. BLS were introduced in Chapter 1, and several examples of physical systems modeled accurately by BLS or combinations thereof were studied in Chapter 2. Numerous applications of BLS models are given in Table 2.1.

In Chapter 3 of Vol. 1, perturbation theory was introduced primarily with regard to linearization about an equilibrium state or a general reference trajectory in order to apply linear-system theory. As shown there, the bilinear term can be included with the linear part of the Taylor series for a more accurate approximation to more general nonlinear systems.

In Section 3.2, BLS stability is studied by means of a quadratic Liapunov function. Here a special class of BLS is introduced which is called dyadic. General concepts of input–output stability for nonlinear systems are briefly summarized. Then stability is used in Chapter 4 in the study of controllability. Observers, structural properties, and solutions are also studied in Chapter 4.

Recall that bilinear systems (BLS) are given by a state differential equation that is linear in state, linear in control, but not jointly linear in both. In other words, according to (1.1) and Fig. 1.1,

$$\frac{d\mathbf{x}}{dt} = \mathbf{A}\mathbf{x} + \sum_{k=1}^{m} \mathbf{N}_k u_k \mathbf{x} + \mathbf{B}\mathbf{u}, \tag{3.1}$$

where $\mathbf{x} \in \mathbb{R}^n$, $\mathbf{u} \in \mathbb{R}^m$; $\mathbf{A}$, $\mathbf{N}_k$, $k = 1, \ldots, m$ and $\mathbf{B}$ are appropriate matrices. Sometimes, the output is given by a linear combination of state variables with

$$\mathbf{y} = \mathbf{C}\mathbf{x}. \tag{3.2}$$

Usually, we assume that the matrices $\mathbf{A}$, $\mathbf{N}_k$, $\mathbf{B}$, and $\mathbf{C}$ are constant so that (3.1) is a time-invariant BLS. For an a priori specified $\mathbf{u}(t)$, BLS (3.1) is a linear time-variant system. If the system has a control feedback of states, (3.1) becomes a more highly nonlinear system. If certain states are measured directly and fed back to generate a feedback control (possibly nonlinear feedback function of states) the system is said to be conditionally linear. Again, conditionally linear systems (CLS) also form an important class of physically meaningful models—as noted in Chapters 1 and 2. In general, such systems may be represented by

$$\begin{aligned} \frac{d\mathbf{x}'}{dt} &= \mathbf{A}'\mathbf{x}' + \sum_{k=1}^{m} \mathbf{N}_k' u_k(\mathbf{y})\mathbf{x}' + \mathbf{B}\mathbf{u}(\mathbf{y}), \\ \frac{d\mathbf{y}}{dt} &= [\mathbf{C} + \mathbf{F}(\mathbf{y})]\mathbf{x}' + \mathbf{g}(\mathbf{y}), \end{aligned} \tag{3.3}$$

where

$$\mathbf{x} = \begin{bmatrix} \mathbf{x}' \\ \mathbf{y} \end{bmatrix},$$

$\mathbf{x}' \in \mathbb{R}^{n-p}$, $\mathbf{y} \in \mathbb{R}^p$; $\mathbf{A}'(\cdot)$, $\mathbf{N}_k'(\cdot)$, $k = 1, \ldots, m$, $\mathbf{B}'$, $\mathbf{C}$, $\mathbf{F}(\cdot)$, and $\mathbf{g}(\cdot)$ are of appropriate dimensions. Note that Eq. (1.2) is merely another form of (3.3).

If (3.1) and (3.3) have a scalar control and scalar output ($m = p = 1$), it is seen that the Luré regulator [Fig. 7.1 and Eq. (7.1) in Vol. 1] becomes a special case of the CLS or the BLS with output feedback. Similarly, the Van der Pol equation (5.25) (and its generalization, Lienard's equation) in Vol. 1 become special cases with appropriate $u(\mathbf{x})$.

3.2 STABILITY

3.2.1 Quadratic Feedback Control

BLS stabilizability is studied in [K6]. It can be shown that the bilinear system (3.1) is stabilizable under certain conditions with a quadratic feedback control $\mathbf{u}(\mathbf{x})$ ([J1], [L2], and [M19]). The presentation here follows that of [G1].

First, system (3.1) is said to be *null stabilizable* if for every $\mathbf{x}(t_0) \in D$ (containing the origin) and every neighbor η of the origin, there exists a locally bounded $\mathbf{u}(t)$, $t \geq t_0$, and a finite time interval T such that the resulting solution $\mathbf{x}(t) \in \eta$ for all $t \geq t_0 + T$. Here D is called a null stabilizable region, and it is assumed that the solution $\mathbf{x}(t)$ to (3.1) is unique and continuous. If $D = \mathbb{R}^n$, the system is stabilizable. Note the similarity to our definition of stability of the origin in terms of δ (in place of D) and ϵ (in place of η) which is developed in Vol. 1, Section 3.3.

By means of a quadratic Liapunov function, $V = \mathbf{x}^T\mathbf{Q}\mathbf{x}$, it can be shown that the control

$$\mathbf{u}_k = -\alpha[\mathbf{N}_k\mathbf{x} + \mathbf{b}_k]^T\mathbf{Q}\mathbf{x}, \qquad \mathrm{k} = 1, \ldots, \mathrm{m}, \tag{3.4}$$

will stabilize (3.1). Here $\alpha > 0$, $\mathbf{Q}$ is positive definite, and $\mathbf{b}_k$ is the kth column of the matrix $\mathbf{B}$.

The proof of this result may be reasoned from the time rate of change of the Liapunov function for (3.1), which is given by

$$\dot{V} = \mathbf{x}^T[\mathbf{QA} + \mathbf{A}^T\mathbf{Q}]\mathbf{x} - 2\alpha \sum_{k=1}^{m} [\mathbf{x}^T\mathbf{Q}(\mathbf{N}_k\mathbf{x} + \mathbf{b}_k)]^2, \tag{3.5}$$

where it is assumed that

$$\begin{bmatrix} (\mathbf{N}_1\mathbf{x} + \mathbf{b}_1)^T\mathbf{Q}\mathbf{x} \\ \vdots \\ (\mathbf{N}_m\mathbf{x} + \mathbf{b}_m)^T\mathbf{Q}\mathbf{x} \end{bmatrix} \neq \mathbf{0} \tag{3.6}$$

for all $\mathbf{x}$ such that $\mathbf{x} \neq \mathbf{0}$ and $\mathbf{x}^T(\mathbf{QA} + \mathbf{A}^T\mathbf{Q})\mathbf{x} \geq 0$. The region D is selected with $V \leq V_m$ such that $V > 0$ and $\dot{V} < 0$ for all $\mathbf{x} \in \eta$. V_m is chosen so that $\mathbf{x}(t_0) \in D$ and so that $\dot{V} < 0$ on the boundary of D. The neighborhood η is defined by the set of $\mathbf{x}$ with $\|\mathbf{x}\| < \epsilon$ such that there exists a constant $\alpha >$ constant $\cdot\, \epsilon^{-2}$. Then, from (3.5) along with a continuous unique $\mathbf{x}(t)$, it follows by repeated convergence arguments over finite time intervals that $\mathbf{x}(t)$ converges to η, a small neighborhood of the origin and remains there. Thus the system is null stabilizable. It is noted that the **A** matrix may have right-hand-plane eigenvalues and still be stabilizable by (3.4) with an appropriate positive α.

Example 3.1 [G1]

Given $\dot{\mathbf{x}} = \mathbf{f}(\mathbf{x}, \mathbf{u})$ such that

$$\begin{aligned} \dot{x}_1 &= x_1 + u_1 x_2 - u_2 x_1, \\ \dot{x}_2 &= x_2 + u_1 x_1 + u_2 x_2. \end{aligned} \tag{3.7}$$

Let $\mathbf{Q} = \mathbf{I}$ so that $V = x_1^2 + x_2^2$, and from (3.4)

$$\begin{aligned} u_1 &= -2\alpha\,(x_1 x_2), \\ u_2 &= \alpha\,(x_1^2 - x_2^2). \end{aligned} \tag{3.8}$$

Then

$$\begin{aligned} \dot{V} &= \frac{\partial V}{\partial \mathbf{x}} \cdot \mathbf{f}(\mathbf{x}, \mathbf{u}) \\ &= 2[x_1 \quad x_2]\begin{bmatrix} x_1 - 2\alpha x_1 x_2^2 - \alpha x_1(x_1^2 - x_2^2) \\ x_2 - 2\alpha x_1^2 x_2 + \alpha x_2(x_1^2 - x_2^2) \end{bmatrix} \\ &= 2[x_1^2 - x_2^2 - \alpha\,(x_1^2 + x_2^2)^2]. \end{aligned} \tag{3.9}$$

It can be seen that $\mathbf{x}(t)$ is forced into η, which is given by the set of $\mathbf{x}$ such that $x_1^2 + x_2^2 < \epsilon$ if $\alpha > 1/\epsilon^2$. While the quadratic feedback control stabilizes this unstable open-loop system very nicely, other combinations are possible. No constant control, however, will do the job.

Example 3.2 [G1]

Consider

$$\begin{aligned} \dot{x}_1 &= x_1/6 + x_2 + (5x_1 + 2x_2)u_1 + (4x_1 + 5x_2)u_2 - 3u_1 - 2u_2, \\ \dot{x}_2 &= x_2/6 + (2x_1 + x_2)u_1 + (5x_1 + 4x_2)u_2 + 2u_1 - 2u_2. \end{aligned} \tag{3.10}$$

Let $\mathbf{Q} = \mathbf{I}$, and note that (3.6) is violated only for $\mathbf{x} = \mathbf{0}$, and $\mathbf{x}^{\mathrm{T}} \approx [-1.6 \quad 4.0]$, where $\mathbf{x}^{\mathrm{T}}(\mathbf{QA} + \mathbf{AQ})\mathbf{x} < 0$. From (3.4), the stabilizing feedback control is given by

$$\begin{aligned} u_1 &= -\alpha\,(5x_1^2 + x_2^2 + 4x_1 x_2 - 3x_1 - 2x_2), \\ u_2 &= -2\alpha\,(2x_1^2 + 2x_2^2 + 5x_1 x_2 + x_1 - x_2). \end{aligned} \tag{3.11}$$

Again, no constant control will stabilize (3.10), but the linear portion of the system is readily stabilized by an appropriate negative linear feedback of the state, $\mathbf{u} = \mathbf{Kx}$.

Now, consider the case of scalar control for the homogeneous BLS (3.1) so that

$$\frac{d\mathbf{x}}{dt} = \mathbf{A}\mathbf{x} + \mathbf{N}\mathbf{x}u, \tag{3.12}$$

where $\mathbf{x} \in \mathbb{R}^n$, $u \in \mathbb{R}$, and $\mathbf{A}$, $\mathbf{N}$ are appropriate constant matrices. Again, assume $V = \mathbf{x}^T\mathbf{Q}\mathbf{x}$ and quadratic feedback control. Further, assume that $\mathbf{Q}$ is selected so that

1. $\mathbf{Q}\mathbf{A} + \mathbf{A}^T\mathbf{Q} = \mathbf{0}$, and
2. the only $\mathbf{x}(0) = \mathbf{x}_0$ for which

$$\mathbf{x}_0^T e^{\mathbf{A}^T t}\mathbf{Q}\mathbf{N}e^{\mathbf{A}t}\mathbf{x}_0 \equiv 0$$

is $\mathbf{x}_0 = \mathbf{0}$ for all t.

Then with [as in (3.4) with $\alpha = 1$]

$$u = -\mathbf{x}^T\mathbf{Q}\mathbf{N}\mathbf{x}, \tag{3.13}$$

it can be shown that (3.12) is globally *asymptotically stable* [Q1]. The proof is provided by [Q1]. Obviously, there is a trade-off of utility between the two results, with the latter providing a stronger result but for a very limited class of systems. Note that condition (1) requires that eigenvalues of $\mathbf{A}$ have zero real part. It also is noted that the Lienard equation and therefore also the Van der Pol equation (5.25) in Vol. 1 are special cases of (3.12) and (3.13).

From (3.12) and (3.13), with $V = \mathbf{x}^T\mathbf{Q}\mathbf{x}$,

$$\dot{V}(\mathbf{x}(t)) = -(\mathbf{x}^T\mathbf{Q}\mathbf{N}\mathbf{x})^2 = -u^2. \tag{3.14}$$

Then, integration of (3.14) over a finite time interval along with a suitable analysis of the boundedness of the evolving positive monotone nonincreasing sequence of Liapunov functions over successive time intervals shows convergence of $V(\cdot)$ to the origin such that

$$V(\mathbf{x}(t)) \leq \frac{c_1}{c_2 t + 1} \tag{3.15}$$

for all $t \geq 0$, where c_1 and c_2 are constants. Then, (3.15) provides a lower bound on the change in $V(\cdot)$ over some duration t.

Requirement 2 can be relaxed somewhat by the addition of a linear term **Ex** to (3.12). Then, following [Q1],

$$\frac{d\mathbf{x}}{dt} = (\mathbf{A} + \mathbf{E})\mathbf{x} + \mathbf{N}\mathbf{x}u, \tag{3.16}$$

where **A**, **N**, and **Q** satisfy the previous hypotheses, and assume that there exists a constant c such that

$$|\mathbf{x}^{\mathrm{T}}\mathbf{Q}\mathbf{E}\mathbf{x}| \le c|\mathbf{x}^{\mathrm{T}}\mathbf{Q}\mathbf{N}\mathbf{x}|,$$
$$\|\mathbf{E}\mathbf{x}\| \le c\|\mathbf{N}\mathbf{x}\|,$$
$$\|\mathbf{N}\mathbf{x}\|^2 \le c\|\mathbf{x}^{\mathrm{T}}\mathbf{Q}\mathbf{N}\mathbf{x}\|,$$

for all $\mathbf{x} \in \mathbb{R}^n$. Then it can be shown that the quadratic feedback control from (3.13) results in an *ultimately bounded solution.*

Example 3.3 [Q1].

Consider a specific Lienard equation in canonical phase variable form

$$\dot{x}_1 = x_2,$$
$$\dot{x}_2 = -x_1 - ux_1,$$

where $u = -\mathbf{x}^{\mathrm{T}}\mathbf{Q}\mathbf{N}\mathbf{x} = x_1x_2$. Consequently, the system is globally asymptotically stable since it can be shown that conditions 1 and 2 are satisfied with $\mathbf{Q} = \mathbf{I}$.

Example 3.4 [Q1].

Given the system

$$\dot{x}_1 = x_2 + x_1 + ux_1, \tag{3.17}$$
$$\dot{x}_2 = -x_1,$$

let

$$\mathbf{Q} = \begin{bmatrix} \frac{1}{3} & 0 \\ 0 & \frac{1}{3} \end{bmatrix}$$

and **A**, **N**, **Q**, and **E** satisfy the required conditions that are given above. Then the quadratic feedback control (3.13) becomes

$$u = -\frac{x_1^2}{3}$$

which yields an ultimately bounded closed-loop system. With this control it is seen that (3.17) is really a generalized Van der Pol equation (5.25) in Vol. 1 with $\mu = 1$.

jor role. For nonlinear systems and bilinear systems in particular, however, singular solutions can be quite significant. Also, it can be shown from (3.21) that if

$$|\mathbf{b}^T\mathbf{PAx}| < |\mathbf{b}^T\mathbf{P}(\mathbf{b} + \mathbf{Nx})| \tag{3.27}$$

in some neighborhood S^* of $\mathbf{x}_s \in \mathcal{S}$, then a trajectory in S^* reaching $\mathcal{S}$ will stay on $\mathcal{S}$ as long as the inequality (3.27) is satisfied. On this switching hyperplane $\mathcal{S}$, the switching function $s(\mathbf{x}) \equiv 0$. That is, $s(\cdot) = \dot{s}(\cdot) = \ddot{s}(\cdot) = \cdots = 0$, where

$$s(\mathbf{x}) = \mathbf{b}^T\mathbf{Px}. \tag{3.28}$$

Then, from (3.21), the so-called sliding control that holds this hyperplane $\mathcal{S}$ is given by

$$u = \psi(\mathbf{x}) = \frac{\mathbf{b}^T\mathbf{PAx}}{\mathbf{b}^T\mathbf{P}(\mathbf{b} + \mathbf{Nx})} \qquad \text{for } s \equiv 0. \tag{3.29}$$

For any physical implementation, however, there is a finite switching time so that the sliding control actually becomes a so-called chattering control which consists of multiple switchings about the switching hyperplane in $\mathbb{R}^n$. Such a policy is investigated along with optimal control policies in Chapter 6. Indeed, the chattering control, if admissible for the application of interest, can provide a very nearly optimal policy.

Assume that

$$V(\mathbf{x}) = \mathbf{x}^T\mathbf{Qx} + s(\mathbf{x})u(\mathbf{x}), \tag{3.30}$$

where $\mathbf{Q}$ is a symmetric positive-definite matrix. Then, if $\mathbf{x} \notin \mathcal{S}$, it is seen from (3.21) that

$$\begin{aligned}\dot{V}(\mathbf{x}) = {} & \mathbf{x}^T(\mathbf{A}^T\mathbf{Q} + \mathbf{QA})\mathbf{x} - \mathbf{x}^T(\mathbf{N}^T\mathbf{Q} + \mathbf{QN})\mathbf{x}u(\mathbf{x}) \\ & - \mathbf{b}^T\mathbf{PNx} + \mathbf{b}^T\mathbf{PAx}u(\mathbf{x}) - 2\mathbf{x}^T\mathbf{Pb}u(\mathbf{x}) - \mathbf{b}^T\mathbf{Pb}.\end{aligned}$$

Next it can be shown [L3] that

$$\dot{V}(\mathbf{x}) \leq k_1\|\mathbf{x}\|^2 + k_2\|\mathbf{x}\| - \mathbf{b}^T\mathbf{Pb},$$

where k_1 and k_2 are positive constants. Then, with $\mathbf{b} \neq \mathbf{0}$ and $\mathbf{P}$ positive definite,

$$\dot{V}(\mathbf{x}) < 0 \tag{3.31}$$

in a neighborhood of the origin. So that with V positive definite, the origin is asymptotically stable.

Next, consider the stability of the sliding motion for (3.21) with sliding control (3.29). In [L3] a Gram–Schmidt orthonormalization of the basis for the switching hyperplane $\mathcal{S}(\mathbf{x})$ is used for this purpose. Assume an orthonormal basis $\{\mathbf{v}_1, \mathbf{v}_2, \ldots, \mathbf{v}_{n-1}\}$ for $\mathcal{S}$ with $\mathbf{v}_n$ orthogonal to $\mathcal{S}$ such that $\|\mathbf{v}_n\| = 1$. Then

$$\mathbf{v}_n = \frac{\mathbf{Pb}}{\|\mathbf{Pb}\|}$$

and $\{\mathbf{v}_1, \ldots, \mathbf{v}_{n-1}, \mathbf{v}_n\}$ forms an orthonormal basis for $\mathbb{R}^n$. The $n \times n$ real transformation matrix $\mathbf{G}$ from the standard orthonormal basis $\{\mathbf{e}_1, \mathbf{e}_2, \ldots, \mathbf{e}_n\}$ of $\mathbb{R}^n$ to the new basis $\{\mathbf{v}_1, \mathbf{v}_2, \ldots, \mathbf{v}_n\}$ is orthogonal. Then with $\mathbf{x} = \mathbf{Gy}$ and $u(\mathbf{x})$ from (3.29), the BLS (3.21) with sliding feedback control becomes

$$\dot{\mathbf{y}} = \mathbf{G}^{\mathrm{T}}\mathbf{AGy} - \mathbf{G}^{\mathrm{T}}(\mathbf{b} + \mathbf{NGy})\frac{\mathbf{b}^{\mathrm{T}}\mathbf{PAGy}}{\mathbf{b}^{\mathrm{T}}\mathbf{P}(\mathbf{b} + \mathbf{NGy})} \tag{3.32}$$

in terms of the new basis. But since $\dot{y}_n = y_n = 0$, let $\mathbf{z} \in \mathbb{R}^{n-1}$ such that $\mathbf{z}^{\mathrm{T}} = [\,y_1\ y_2\ \ldots\ y_{n-1}]$. Then (3.32) becomes

$$\dot{\mathbf{z}} = \mathbf{H}^{\mathrm{T}}\mathbf{AHz} - \mathbf{H}^{\mathrm{T}}(\mathbf{b} + \mathbf{NHz})\frac{\mathbf{b}^{\mathrm{T}}\mathbf{PAHz}}{\mathbf{b}^{\mathrm{T}}\mathbf{P}(\mathbf{b} + \mathbf{NHz})} = \mathbf{f}(\mathbf{z}), \tag{3.33}$$

where $\mathbf{H} \in \mathbb{R}^{n \times (n-1)}$ is identical to $\mathbf{G}$ with the nth column removed, and $\mathbf{f}(\mathbf{0}) = \mathbf{0}$. Now, let

$$\mathbf{F} = \left[\frac{\partial \mathbf{f}}{\partial \mathbf{z}}(\mathbf{z} = \mathbf{0})\right] \tag{3.34}$$

be the Jacobian matrix of $\mathbf{f}(\cdot)$ at $\mathbf{z} = \mathbf{0}$.

Then it is shown in [L3] that the BLS (3.21) with $u(\cdot)$ from (3.23), including sliding control (3.29), has an asymptotically stable origin if $\mathbf{F}$ is Hurwitz (i.e., all the eigenvalues of $\mathbf{F}$ have negative real parts).

Example 3.5 [L3]

Consider the continuous stirred tank reactor with a first-order irreversible exothermic reaction under nonisothermal conditions. Cooling of the reactor is achieved by a fluid coolant. Then the process is approximated by the BLS (3.21) with

$$\mathbf{A} = \begin{bmatrix} 4.25 & 1 \\ -6.25 & -2 \end{bmatrix}, \quad \mathbf{b} = \begin{bmatrix} -0.25 \\ 0 \end{bmatrix}, \quad \mathbf{N} = \begin{bmatrix} -1 & 0 \\ 0 & 0 \end{bmatrix}.$$

Since **A** has a positive eigenvalue, the open-loop linear system is unstable. x_1 and x_2 are changes in output temperature and concentration, respectively, from their equilibrium values. The control u represents the effect of coolant flow rate such that $|u| \leq 1$. All variables are normalized and dimensionless. The control is to be manipulated to maintain the output temperature and chemical reaction near their equilibrium values.

Let $\mathbf{P} = \mathbf{I}$ in (3.24)–(3.29) so that

$$s(\mathbf{x}) = -0.25x_1$$

and

$$u = \begin{cases} +1 & \text{if } x_1 < 0 \\ -1 & \text{if } x_1 > 0 \\ 4x_2 & \text{if } x_1 = 0. \end{cases}$$

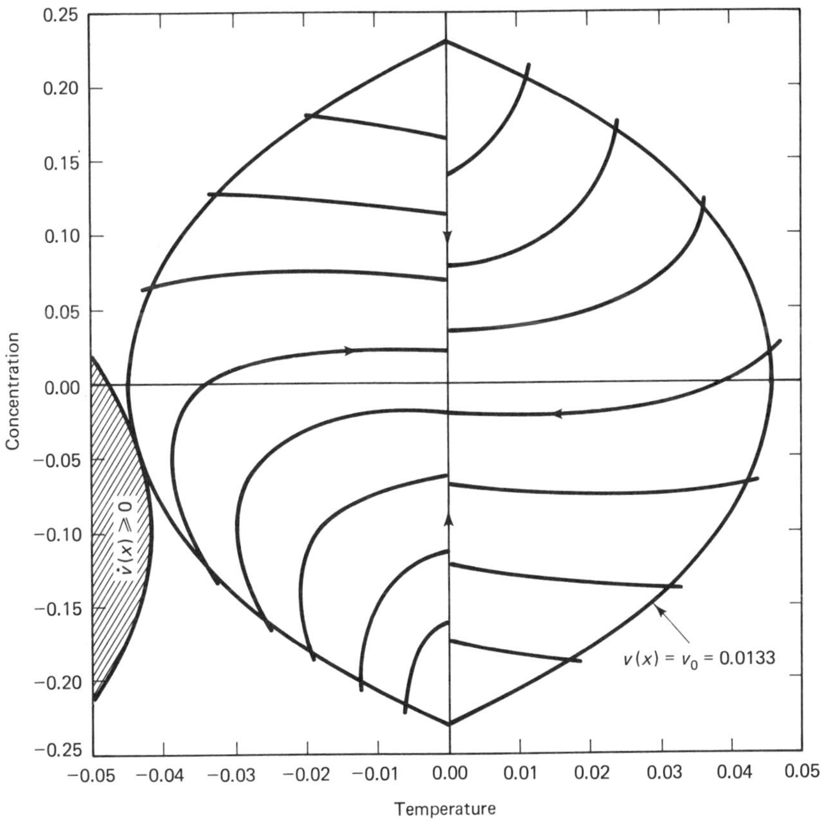

Figure 3.1 Region of asymptotic stability for Example 3.5. (After [L3].)

Then, with the orthonormal basis

$$v_1 = e_2 \qquad \text{and} \qquad v_2 = -e_1$$

as explained above,

$$\mathbf{G} = \begin{bmatrix} 0 & -1 \\ 1 & 0 \end{bmatrix},$$

$\mathbf{H}^{\mathrm{T}} = [0 \quad 1]$, and from (3.33),

$$\dot{\mathbf{z}} = -2\mathbf{z}.$$

Consequently, the origin is asymptotically stable, with

$$\mathbf{Q} = \begin{bmatrix} 1 & 0 \\ 0 & 0.25 \end{bmatrix}.$$

Figure 3.1 shows the resulting state-plane trajectories and the guaranteed region of asymptotic stability bounded by $v(\mathbf{x}) = v_m = 0.0133$ with $\dot{V}(x) < 0$ interior to the region enclosed except at $\mathbf{0}$.

3.2.4 Dyadic Bilinear Systems

The bilinear system (3.1) is said to be *dyadic of order d* if for $k = k_1, \ldots, k_d$, where $d \in \{1, 2, \ldots, m\}$,

$$\mathbf{N}_k\mathbf{x} + \mathbf{b}_k = \mathbf{b}_k(\mathbf{c}_k^{\mathrm{T}}\mathbf{x} + 1). \tag{3.35}$$

Here $\mathbf{c}_k^{\mathrm{T}}$ is a constant m-dimensional row vector. Many physical systems can be modeled by such dyadic BLS. For example, consider the flow system of Fig. 3.2, which models the transfer of some dissolved living matter between compartments in the human body [G1]. A conservation (or continuity) of mass yields

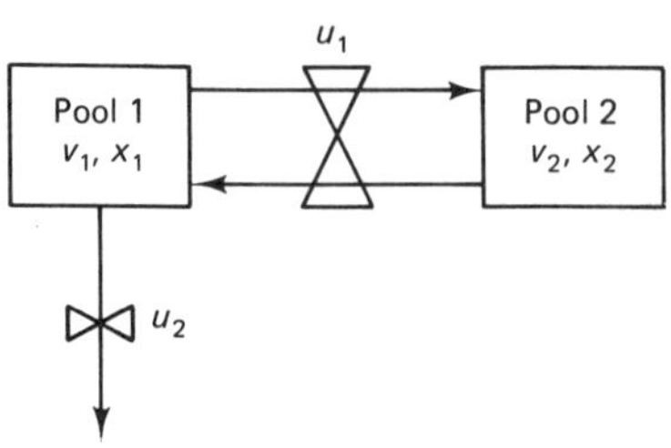

Figure 3.2 Human-body dyadic flow system.

$$\begin{aligned} v_1\dot{x}_1 &= u_1(x_1 - x_2) + u_2x_1, \\ v_2\dot{x}_2 &= -u_1(x_1 - x_2) + \alpha_2x_2, \end{aligned} \tag{3.36}$$

where v_1, v_2 = volume of compartment 1, 2
x_1, x_2 = concentration of compartment 1, 2
u_1 = transfer rate between compartments
u_2 = transfer rate from compartment 1 (not to compartment 1)
α_2 = growth rate of matter in compartment 2

u_1, u_2, x_1, and x_2 are all nonnegatives. In the form of BLS (3.1), (3.36) becomes

$$\frac{d\mathbf{x}}{dt} = \mathbf{A}\mathbf{x} + \mathbf{N}_1u_1\mathbf{x} + \mathbf{N}_2u_2\mathbf{x}, \tag{3.37}$$

where

$$\mathbf{A} = \begin{bmatrix} 0 & 0 \\ 0 & \dfrac{\alpha_2}{v_2} \end{bmatrix}, \qquad \mathbf{N}_1 = \mathbf{n}_1\mathbf{d}_1^T = \begin{bmatrix} \dfrac{1}{v_1} \\ -\dfrac{1}{v_2} \end{bmatrix} [1 \quad -1], \qquad \mathbf{N}_2 = \mathbf{n}_2\mathbf{d}_2^T = \begin{bmatrix} \dfrac{1}{v_1} \\ 0 \end{bmatrix} [1 \quad 0].$$

$\mathbf{N}_1$ and $\mathbf{N}_2$ are called dyads.

The equilibrium state $\mathbf{x}_e^T = [x_{1e} \quad x_{2e}]$, with x_{1e} and x_{2e} positive, requires from (3.37) that

$$\mathbf{u}_e = \begin{bmatrix} \dfrac{\alpha_2x_{2e}}{x_{1e} - x_{2e}} \\ \dfrac{-\alpha_2x_{2e}}{x_{1e}} \end{bmatrix}.$$

Let $\mathbf{y} = \mathbf{x} - \mathbf{x}_e$ and $\mathbf{v} = \mathbf{u} - \mathbf{u}_e$ to transform $\mathbf{x}_e$ and $\mathbf{u}_e$ to the origin. Then

$$\frac{d\mathbf{y}}{dt} = \mathbf{A}'\mathbf{y} + \mathbf{n}_1'(\mathbf{d}_1'^{\mathrm{T}}\mathbf{y} + 1)v_1 + \mathbf{n}_2'(\mathbf{d}_2'^{\mathrm{T}}\mathbf{y})v_2 \tag{3.38}$$

is dyadic of order 2.

Now consider the scalar input dyadic system

$$\frac{d\mathbf{x}}{dt} = \mathbf{A}\mathbf{x} + [\mathbf{N}\mathbf{x} + \mathbf{b}]u, \tag{3.39}$$

where

$$[\mathbf{N}\mathbf{x} + \mathrm{b}] = \mathbf{b}[\mathbf{c}^{\mathrm{T}}\mathbf{x} + 1].$$

Let $d(\mathbf{x}) = \mathbf{c}^{\mathrm{T}}\mathbf{x} + 1$, so that $d(\mathbf{x}) = 0$ defines an $(n - 1)$-dimensional hyperplane. The hyperplane divides the state space into the following three sets [similar to that of (3.24)–(3.26)]:

$$\begin{aligned}\mathcal{S}_+ &= \{\mathbf{x} : d(\mathbf{x}) > 0\},\\ \mathcal{S}_- &= \{\mathbf{x} : d(\mathbf{x}) < 0\},\\ \mathcal{S} &= \{\mathbf{x} : d(\mathbf{x}) = 0\}.\end{aligned}$$

Now, by means of an appropriate variable-structure division controller, (3.39) becomes linear with piecewise-constant coefficients. For example, in [G1] it is proposed to use

$$u = \begin{cases} \dfrac{\mathbf{k}_+^{\mathrm{T}}\mathbf{x} + u_r(\mathbf{x}, t)}{d(\mathbf{x})}, & \mathbf{x} \in \mathcal{S}'_+ \\ \dfrac{\mathbf{k}_-^{\mathrm{T}}\mathbf{x} + u_r(\mathbf{x}, t)}{d(\mathbf{x})}, & \mathbf{x} \in \mathcal{S}'_- \\ 0, & \mathbf{x} \in \mathcal{S}', \end{cases} \tag{3.40}$$

where

$$\mathcal{S}' = \{\mathbf{x} : -\epsilon_1 \le d(\mathbf{x}) \le \epsilon_2\},$$

where ϵ_1 and ϵ_2 are nonnegative constants. $\mathcal{S}'$ is removed from $\mathcal{S}_+$ and $\mathcal{S}_-$ to form $\mathcal{S}'_+$ and $\mathcal{S}'_-$, respectively. $u_r(\mathbf{x}, t)$ is a reference control in order to shift equilibrium points. As will be shown in Section 4.2, an additive control is important to control the BLS to different equilibrium points. Note that dyadic (3.39) as it stands does not exhibit additive control. k_1, k_2, ϵ_1, and ϵ_2 are selected appropriately to stabilize the system such as by the existence of a quadratic Liapunov function for the linear piecewise-constant closed-loop system

$$\begin{aligned}\frac{d\mathbf{x}}{dt} &= [\mathbf{A} + \mathbf{b}\mathbf{k}_+^{\mathrm{T}}]\mathbf{x} + \mathbf{b}u_r, && \mathbf{x} \in \mathcal{S}'_+,\\ \frac{d\mathbf{x}}{dt} &= [\mathbf{A} + \mathbf{b}\mathbf{k}_-^{\mathrm{T}}]\mathbf{x} + \mathbf{b}u_r, && \mathbf{x} \in \mathcal{S}'_-,\\ \frac{d\mathbf{x}}{dt} &= \mathbf{A}\mathbf{x}, && \mathbf{x} \in \mathcal{S}'\end{aligned} \tag{3.41}$$

Note that it is a relatively simple matter to stabilize the system by appropriate selection of $\mathbf{k}_+$ and $\mathbf{k}_-$ once it is known to be stabilizable.

Example 3.6 [G1]

Consider the nuclear fission process Vol. 1, (2.122) and (2.123), in a normalized form with

$$x_1 = \frac{n - n_e}{n_e} > -1,$$

$$x_2 = \frac{c - c_e}{c_e} > -1.$$

Then

$$\dot{\mathbf{x}} = \mathbf{A}\mathbf{x} + \mathbf{b}d(\mathbf{x})u, \tag{3.42}$$

where

$$\mathbf{A} = \begin{bmatrix} -\dfrac{\beta}{l} & \dfrac{\beta}{l} \\ \lambda & -\lambda \end{bmatrix}, \qquad d(\mathbf{x}) = \frac{x_1 + 1}{l}, \qquad \mathbf{b} = \begin{bmatrix} 1 \\ 0 \end{bmatrix}.$$

The eigenvalues of $\mathbf{A}$ are $\lambda_1 = 0$ and $\lambda_2 = -(\lambda + \beta/l)$. Note that by this change of equilibrium state to the origin the nuclear fission process becomes dyadic.

For this process $\mathcal{S} = \{\mathbf{x} : d(\mathbf{x}) = 0\}$ does not belong to the admissible state space. Since $[\mathbf{A}, \mathbf{b}]$ is a controllable pair, it is relatively easy to select $\mathbf{k}$ to asymptotically stabilize the system (3.42) with the division controller

$$u = \frac{\mathbf{k}^T\mathbf{x}}{d(\mathbf{x})}.$$

In this case, (3.42) becomes the equivalent linear system

$$\dot{\mathbf{x}} = \mathbf{A}\mathbf{x} + \mathbf{b}u \tag{3.43}$$

with control $u = \mathbf{k}^T\mathbf{x}$. For example, $\mathbf{k}^T = [648.4 \quad -650.9]$ proved to be asymptotically stabilizing with excellent response characteristics.

The asymptotically stabilizing quadratic control by (3.4) becomes

$$u = -[d(\mathbf{x}) \quad 0]\mathbf{Q}\mathbf{x}$$

with Liapunov function $V = \mathbf{x}^T\mathbf{Q}\mathbf{x}$. Here the positive-definite $\mathbf{Q}$ is given by

$$\mathbf{Q} = \begin{bmatrix} q_{11} & q_{12} \\ q_{12} & q_{22} \end{bmatrix},$$

so that

$$u = -\frac{x_1 + 1}{l}(q_{11}x_1 + q_{12}x_2).$$

$\mathbf{Q}$ may be selected with $q_{11} = 40(10)^{-10}$, $q_{12} = 20(10)^{-10}$, and $q_{22} > q_{12}^2/q_{11}$. However, the response with this controller was slower than the division controller above. Of course, they could both be improved on with further analysis and computer simulation.

It should be cautioned that cancellation of nonlinear dynamics generally is even riskier than pole–zero cancellation for linear systems. Consequently, the performance of the division controller may not be very robust with respect to unknowns in the critical plant parameters or even the structure itself. The designer should realize that a plant model is only some approximation to the real system. This is in addition to inaccuracies in measurements and in the control synthesis. Nevertheless, the literature is rich with reference to "making nonlinear systems linear" in order to use traditional linear design methodologies. Also, it will be seen that nonlinear systems may very well exhibit superior performance than their linear system counterparts. For example, see [S3] and Section 5.4, Vol. 1.

While the theory of nonlinear regulator system stability is fairly rich as studied thus far, the general theory for forced-system stability is not nearly so well developed. Of course for linear systems, there is a nice correlation between bounded-input, bounded-output (BIBO) stability and null stability, as discussed in Vol. 1, Chapter 3. An excellent summary of input–output stability for nonlinear systems is included in [V2]. Pioneering research in this area is included in [Z1]. Since most of this work relies on a knowledge of functional analysis, it will not be studied in this book.

For the BLS (3.21), however, it is not too difficult to show that if $\mathbf{A}$ is a stable matrix and

$$\int_0^\infty |u(\sigma)|\, d\sigma < \infty, \tag{3.44}$$

then $\mathbf{x}(t)$ is bounded for $t \in [0, \infty]$, where u is a scalar control. If $\mathbf{A}$ is Hurwitz and there exists a nonsingular transformation matrix $\mathbf{P}$ such that $\mathbf{PAP}^{-1}$ and $\mathbf{PNP}^{-1}$ are triangular, then $\mathbf{x}(t)$ is uniformly bounded on $[0, \infty)$ if (3.44) is satisfied.

EXERCISES

3.1. Study the stability of the origin when using a sliding control policy along a switching manifold given by $\mathbf{P} = \mathbf{I}$[i.e., $S(\mathbf{x}) = \mathbf{b}^T\mathbf{P}\mathbf{x}$] for the following system, as a function of a and d:

$$\dot{x}_1 = x_2,$$

$$\dot{x}_2 = x_3 + u,$$

$$\dot{x}_3 = ax_1 + dx_2 - 2x_3 + x_2u - u.$$

3.2. Consider the following bilinear system:

$$\dot{x}_1 = x_2,$$

$$\dot{x}_2 = x_1 + x_2 + (a + e)x_2u + u,$$

where the state x_1, x_2 is perfectly measured. $a = 3$ is a parameter corrupted by an error e made during identification.

(a) Is the system dyadic?

(b) Compute a division control such that double poles occur at -3 if $d(x) > 0$ and at -2 if $d(x) < 0$, and the origin is the equilibrium point.

(c) Plot and compare the trajectories for $t \in [0, 2]$ and $(x_1, x_2)_{t=0} = (1, 0)$ for the following three cases: $e = 0$ (error-free model), $e = +0.05$, and $e = -0.05$. When is a division controller an unsatisfactory solution?

3.3. Show how the Popov method (Section 7.2, Vol. 1) can ascertain stability for a class of BLS.

3.4. For some BLS, additive linear state feedback control roughly can be used for local stability and parametric control roughly to supplement stability in the large. Give this statement more precise meaning by means of Liapunov's direct method (Section 6.2.4, Vol. 1).

3.5. Consider the following BLS with linear state feedback:

$$\dot{x}_1 = -x_1 + ux_2,$$

$$\dot{x}_2 = -x_2,$$

where $u = x_2$.

(a) Ascertain a region of stability at $\mathbf{x} = \mathbf{0}$ by Liapunov's direct method.

(b) Compute $x_2(t)$ for $x_1(t_0) = x_{10}$, $x_2(t_0) = x_{20}$.

(c) Compare and discuss the results of parts (a) and (b).

3.6. Consider the feedback control of the following plant:

$$\dot{x}_1 = -3x_1 - 3(x_1 + x_2)u,$$

$$\dot{x}_2 = 0.2x_1 - x_2 - 3x_2u + u.$$

(a) Study the stability of $\mathbf{x} = \mathbf{0}$ if $u = 0.2(x_1 + x_2)$.
(b) Derive a quadratic stabilizing control $u(\mathbf{x})$ for this problem.
(c) Discuss the relative merits of linear and quadratic feedback and of independent additive and multiplicative (parametric) control for this problem.

CHAPTER 4

STRUCTURAL PROPERTIES

4.1 INTRODUCTION

Controllability, its relationship to stability, observer design, and functional combinations of bilinear systems (BLS) are studied in this chapter. Linear system controllability is reviewed briefly in Section 4.2.1. It is shown that asymtotically stable linear constant-coefficient systems with bounded-magnitude constraints are not controllable. In many cases, however, a bilinear control term with similar magnitude constraints may be introduced which will make the system controllable. This is a common feature of biological processes such as the human immune system studied in Chapter 2, where, for example, antibody production is turned-on by helper-T cells (in conjunction with other mechanisms) and is turned-off by suppressor-T cells. Such processes acting in concert are able to control many rapidly reproducing alien invaders. Of course, the potential benefit of such nonlinear control in engineering design is obvious. Sufficient conditions, as well as certain necessary conditions, are derived for BLS controllability in Section 2.2.3.

There is a brief summary of alternative forms of controllability in Section 4.2.3—some of which are more conducive to analysis of nonlinear systems but not necessarily as important in practice as conventional controllability. It is found that Lie algebra can be convenient for certain controllability analyses. A detailed analysis of this approach is beyond the scope of this book. However, a brief introduction is presented in Section 4.2.3 with some background on Lie algebra provided in Appendix A. In Appendix A, the solution to certain BLS are computed. It appears that Lie algebra can reduce the computational steps significantly. However, many readers may wish to skip Section 4.2.3 since the material is not used elsewhere in this book. Those with a desire for more detail should consult the references of that section.

4.2 CONTROLLABILITY

Controllability was introduced in Vol. 1, Section 3.5, where several convenient results were presented for linear systems. There a system was defined to be controllable if any given initial state can be driven to any terminal state in finite time. It is implicitly assumed that the states of concern are finite. For linear time-invariant systems, a convenient rank test (3.93) in Vol. 1 is available to determine controllability. Furthermore, the controllable subspace is spanned by the column vectors of M, the controllability matrix in (3.93), Vol. 1. Also, for linear time-invariant systems, controllability is equivalent to the ability to steer the system from the origin to any other state as a consequence of the superposition property; controllability is independent of the duration time, and local controllability is identical to controllability. For nonlinear systems, however, none of these is generally true. On the other hand, nonlinear systems potentially offer more possibilities of controllability than do linear systems. For example, the BLS (3.1) has the possibility of controllability through the pairs $[\mathbf{A}, \mathbf{B}]$, $[\mathbf{A}, \mathbf{N}_k]$, and $[\mathbf{B}, \mathbf{N}_k]$, $k = 1, \ldots, m$, by appropriate system design.

In the next section, the apparent lack of controllability of asymptotically stable linear systems with bounded control is established. Then in Section 4.2.2, sufficient conditions for controllability of bilinear systems, with magnitude-constrained control, are developed. Finally, other forms of controllability are summarized. The treatment primarily follows [M2], [R3], and [V2].

4.2.1 Linear System Limitation

Controllability, to be more practical and precise, is defined as usual with the added requirement that the states must be reached with an admissible control. That is, a system is *controllable* if it can be forced from any initial state to any terminal state in finite time *with an admissible control*. Typically, the control $\mathbf{u}(t)$ must be at least piecewise continuous and confined to a closed bounded set U. For example, $|u_k(t)| \leq 1$, $k = 1, \ldots, m$.

From the existence of a positive-definite quadratic Liapunov function, it is shown that an asymptotically stable linear system with $\mathbf{u} \in U$ cannot be controllable. Here, again, consider the linear system

$$\frac{d\mathbf{x}}{dt} = \mathbf{Ax} + \mathbf{Bu}, \tag{4.1}$$

where $\mathbf{x} \in \mathbb{R}^n$, $\mathbf{u} \in U$, a closed bounded set in $\mathbb{R}^m$; $\mathbf{A}$ is Hurwitz (i.e., all eigenvalues in the left-half plane); and $\mathbf{A}$ and $\mathbf{B}$ are constant matrices of appropriate dimensions.

Theorem 1. The linear system (4.1) with $\mathbf{u}$ bounded is *not* completely controllable if the eigenvalues of $\mathbf{A}$ have negative real parts.

To show this, first consider the autonomous system

$$\frac{d\mathbf{x}}{dt} = \mathbf{Ax}, \tag{4.2}$$

where $\mathbf{A}$ has negative-real eigenvalues (i.e., $\mathbf{A}$ is Hurwitz). Since (4.2) is asymptotically stable, there exists a positive-definite quadratic form (a Liapunov function) $V(\mathbf{x})$ with $\dot{V}(\mathbf{x})$ a negative-definite quadratic form $p(\mathbf{x})$. Then the surfaces $V(\mathbf{x}) =$ constant are hyperspheroids that enclose the origin in state space.

The same system with an added bounded linear control term is defined by (4.1). By taking the inner product of the gradient of V with the state velocity it is seen that

$$\dot{\mathbf{x}} \cdot \frac{\partial V}{\partial \mathbf{x}} = \mathbf{Ax} \cdot \frac{\partial V}{\partial \mathbf{x}} + \mathbf{Bu} \cdot \frac{\partial V}{\partial \mathbf{x}} = p(\mathbf{x}) + \mathbf{Bu} \cdot \frac{\partial V}{\partial \mathbf{x}}, \tag{4.3}$$

where $p(\mathbf{x})$ is negative definite and the last term is linear in $\mathbf{x}$. Now (4.3) is negative for every bounded $\mathbf{u}$ with $\mathbf{x}$, and therefore $V(\mathbf{x}) = V_M$, sufficiently large. Hence the state velocity vector $\dot{\mathbf{x}}$ points into the interior of the volume, which is bounded by $V(\mathbf{x}) = V_M$, and the stable linear system is not completely controllable with bounded control. Similarly, if (4.2) is unstable, there is a region defined by a quadratic $V(\mathbf{x})$ such that the velocity vectors for (4.1) all point outward. Consequently, the unstable linear time-invariant system is not controllable with bounded control.

This short proof suggests that there is some fundamental relationship between stability and controllability. In Section 4.2.2 this relationship is further exploited to provide sufficient conditions for complete controllability of bilinear systems.

It is of interest to note that the linear harmonic oscillator does not fit the above-mentioned class of linear systems lacking controllability. It is readily seen, for example, that the second-order linear system with imaginary eigenvalues is completely controllable with bounded control if a sufficiently large number of switchings are permitted with a bang-bang control.

4.2.2 Controllability with Bounded Control

The variable structure of bilinear systems allows them to be more controllable than linear systems, just as it frequently provides for a more accurate model. In control system design, it may be convenient or even necessary to introduce adaptive or variable-structure control to steer the plant throughout the required region of state space. Controllability already has been introduced as a useful concept from a mathematical standpoint. While complete controllability can never be realized in practice due to physical constraints, there are numerous pursuit systems, such as the immune system or the ABM, for which it may be necessary to attain relatively large regions of state space. The following example shows how a linear system that is not completely controllable with bounded control may be made so by the addition of a bilinear mode of control.

Example 4.1 [M2]

Consider the system

$$\frac{dx_1}{dt} = x_2,$$
$$\frac{dx_2}{dt} = -2x_1 - x_2 + u. \tag{4.4}$$

First, with unbounded control $u(t)$, it is seen that the rank of the controllability matrix, **M** in Vol. 1, (3.93), is equal to 2, the system order; hence the system is completely controllable if infinite control is allowed. If the control is constrained, however, so that $|u| < 1$, it is readily seen from the theorem above that the system is not controllable. It is apparent from the so-called bang-bang principle [L5] and the characteristic, stable-focus, state behavior for the homogeneous part of (4.4) that the constrained system does have a degree of controllability or null controllability [L5] such that any state between the lines $x_1 = -\frac{1}{2}$ and $x_1 = +\frac{1}{2}$ on the (x_1, x_2) state plane may be driven to the origin in finite time.

If any appropriate bilinear control is added to (4.4), however, the system can be made completely controllable. For example, consider

$$\dot{x}_1 = x_2,$$
$$\dot{x}_2 = -2x_1 - x_2 + u + x_1 u + 2x_2 u, \tag{4.5}$$

with $|u| \leq 1$. For $u = +1$, the system exhibits an unstable-focus characteristic with an equilibrium point at (1, 0) on the phase plane, and for $u = -1$ a stable focus with equilibrium point at $(-\frac{1}{3}, 0)$. The corresponding eigenvalues are $\lambda_1, \lambda_2 = \frac{1}{2}(1 \pm j\sqrt{3})$ and $\lambda_1, \lambda_2 = -(\frac{1}{2})(3 \pm j\sqrt{3})$, respectively. It is apparent that such combinations of stable and unstable foci are completely state controllable. In general, if the bilinear control that is added to the second equation of (4.5) has the form

$$n_{21}x_1 u + n_{22}x_2 u,$$

then an evaluation of the system eigenvalues with $u = +1$ and $u = -1$ shows that a combination of stable and unstable foci (thus ensuring complete controllability) exists if $n_{22} \in [1, \sqrt{7}]$ and

$$\frac{n_{22}^2 + 2n_{22} - 7}{4} < n_{21} < \frac{-n_{22}^2 + 2n_{22} + 7}{4}$$

This example shows that, roughly speaking, additive linear control may be used to control new equilibrium states, while multiplicative linear control may be used to vary the structure and therefore the range of attainable states—a property that is exploited more precisely below. The following example shows that, as should be expected, not every bilinear system is completely controllable.

Example 4.2 [M2]

Again, the linear system

$$\frac{dx_1}{dt} = x_2,$$
$$\frac{dx_2}{dt} = x_1 - x_2 + u \tag{4.6}$$

is obviously completely controllable with infinite control allowed since the rank of **M** from Vol. 1, (3.93), is $n = 2$. Even though the linear system is unstable and does not fit Theorem 1, the system is not completely controllable with bounded control. In this case there is a region of null controllability between two parallel lines corresponding to the stable eigenvectors for the saddle point at $(-1, 0)$ and at $(+1, 0)$ with $u = +1$ and $u = -1$, respectively. If the bilinear control $-x_2u$ is added to the second equation of (4.6), the system still exhibits saddle-point portraits for $u = +1$ and $u = -1$. Again, the superposition of these state-plane portraits will yield an enlarged region of controllability, in this case not the entire state space. Again, however, it is possible to show complete controllability by a superposition of stable- and unstable-foci portraits for $u = +1$ and $u = -1$ with an appropriate bilinear control.

Figure 4.1 shows typical state-plane portraits for bang-bang-controlled second-order BLS. Note the regions of controllability. While multiplicative control by its variable structure is a valuable asset to controllability, the need for additive control is obvious from the previous work. This point is shown vividly by the simple first-order system

$$\dot{x} = ax + bux + cu,$$

where $|u| \leq 1$. It is immediately obvious that this system is completely controllable if $a = 0$, $b = 0$, and $c \neq 0$. By considering the other combinations of coefficients it is just as easy to show that this is a necessary condition as well.

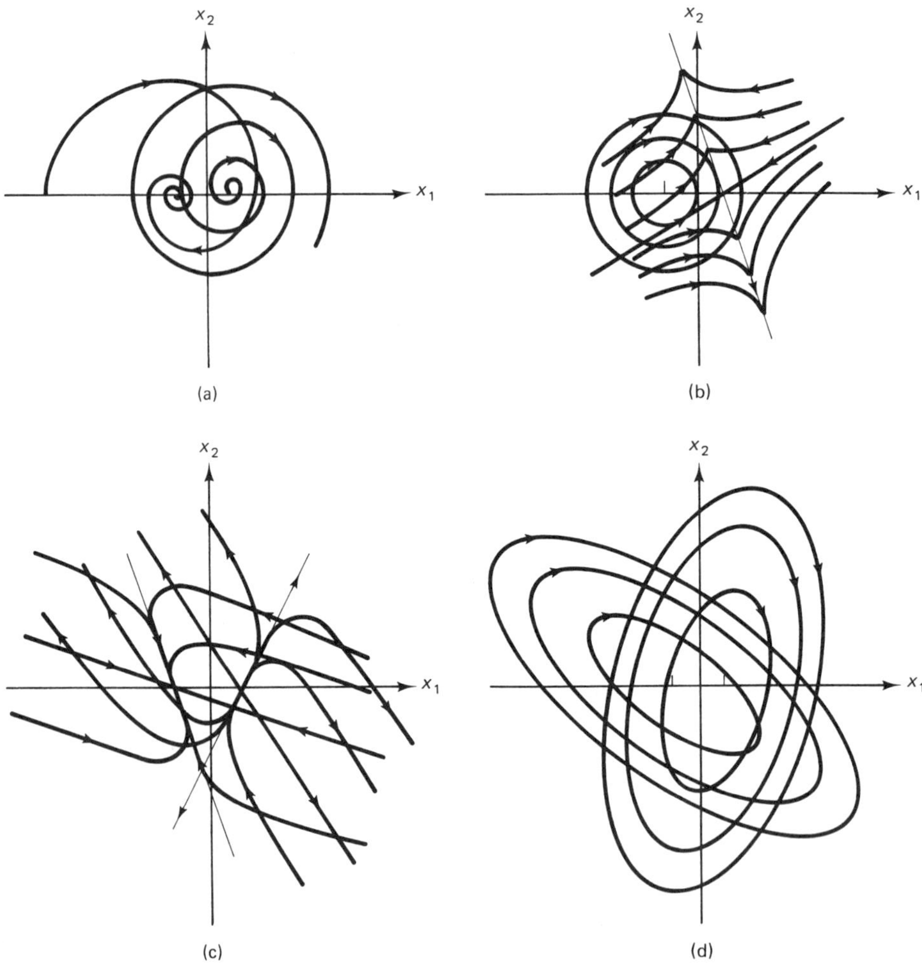

Figure 4.1 Typical bang-bang-controlled BLS state planes: (a) stable–unstable foci, complete controllability; (b) center–saddle points, limited controllability; (c) stable–unstable nodes, limited controllability; (d) center–center points (adaptive pumping oscillator), complete controllability.

On the other hand, similar calculations readily show that the system with two independent controls, such that

$$\dot{x} = ax + bux + cv,$$

where $|u| \leq 1$, $|v| \leq 1$, is completely controllable if and only if $|b| \geq |a|$ and $c \neq 0$.

There are numerous necessary conditions that can be derived to determine the controllability of bilinear systems with unconstrained or constrained control. Some of these are presented here with the proofs left as exercises for the reader.

Necessary conditions for BLS controllability

1. For the bilinear system (3.21) with scalar control to be completely controllable, it is necessary that $\mathbf{A}$ has distinct nonzero eigenvalues and that $\mathbf{A}$ and $\mathbf{N}$ have no common eigenvectors.
2. Define the following sets for BLS (3.1):

$$\begin{aligned}\mathbf{W}_A &= \{\mathbf{x} : \mathbf{A}\mathbf{x} = \mathbf{0}\}\\ \mathbf{W}_1 &= \{\mathbf{x} : \mathbf{N}_1\mathbf{x} = -\mathbf{b}_1\}\\ &\vdots\\ \mathbf{W}_m &= \{\mathbf{x} : \mathbf{N}_m\mathbf{x} = -\mathbf{b}_m\},\end{aligned}$$

where $\mathbf{B} = [\mathbf{b}_1\ \mathbf{b}_2\ \ldots\ \mathbf{b}_m]$. Then for the bilinear system to be completely controllable it is necessary that the intersection of these sets is the null set, or

$$\mathbf{W}_A \cap \mathbf{W}_1 \cdots \cap \mathbf{W}_m = \emptyset.$$

3. A necessary condition that the bilinear system with scalar control (3.21) be completely controllable is that

$$\begin{aligned}\text{rank}\ [&(\mathbf{N}\mathbf{x}_e + \mathbf{b}), (\mathbf{A} + \mathbf{N}u_e)(\mathbf{N}\mathbf{x}_e + \mathbf{b}), \mathbf{N}(\mathbf{N}\mathbf{x}_e + \mathbf{b}),\\ &\ldots, (\mathbf{A} + \mathbf{N}u_e)^{n-1}(\mathbf{N}\mathbf{x}_e + \mathbf{b}),\\ &\ldots, \mathbf{N}^{n-1}(\mathbf{N}\mathbf{x}_e + \mathbf{b})] = n,\end{aligned}$$

at every equilibrium point $\mathbf{x}_e$ with the required equilibrium control u_e. It is relatively easy to generalize this condition for vector control with only a major complication in the algebra.

Example 4.3

$$\dot{\mathbf{x}} = \mathbf{A}\mathbf{x} + \mathbf{N}u\mathbf{x} + \mathbf{b}u, \tag{4.7}$$

where

$$\mathbf{A} = \begin{bmatrix}1 & 0\\ 0 & 0\end{bmatrix}, \quad \mathbf{N} = \begin{bmatrix}1 & 0\\ 0 & 2\end{bmatrix}, \quad \mathbf{b} = \begin{bmatrix}1\\ 1\end{bmatrix}.$$

For condition 2,

$$\mathbf{W}_A = \left\{\mathbf{x} : \mathbf{x} = \begin{bmatrix} 0 \\ x_2 \end{bmatrix}\right\},$$

$$\mathbf{W}_1 = \left\{\mathbf{x} : \mathbf{x} = \begin{bmatrix} -1 \\ -\frac{1}{2} \end{bmatrix}\right\}.$$

Hence $\mathbf{W}_A \cap \mathbf{W}_1 = \emptyset$, as required.

For condition 3, however, at

$$\mathbf{x}_e = \begin{bmatrix} 0 \\ -\frac{1}{2} \end{bmatrix}, \qquad u_e = 0,$$

the rank test yields

$$\text{rank} \begin{bmatrix} 1 & 1 & 1 \\ 0 & 0 & 0 \end{bmatrix} = 1 \neq n.$$

Consequently, the system is not completely controllable despite the satisfaction of condition 2.

Note that if the $\mathbf{N}_k$ matrices vanish, condition 3 reduces to the necessary and sufficient condition so familiar for linear systems.

Sufficient conditions for complete controllability of bilinear systems are derived in [R1] and [M2] and the analysis presented next follows that work. Again the bilinear system is given by

$$\frac{d\mathbf{x}}{dt} = \mathbf{f}(\mathbf{x}, \mathbf{u}) = (\mathbf{A} + \sum_{k=1}^{m} u_k \mathbf{N}_k)\mathbf{x} + \mathbf{B}\mathbf{u}, \tag{4.8}$$

where the parameters are as defined before. Now it is assumed that the class of admissible controls $\{\mathbf{u}(t)\}$ is the class of all piecewise-continuous vector time functions with domain $[0, \infty)$ and range U, where U is a compact connected set containing the origin in $\mathbb{R}^m$.

Again consider the nonlinear system

$$\frac{d\mathbf{x}}{dt} = \mathbf{f}(\mathbf{x}, \mathbf{u}), \tag{4.9}$$

where $\mathbf{f}(\cdot)$ is continuously differentiable in its arguments. Then consider the linearized representation of (4.9) about $\mathbf{f}(\mathbf{x}, \mathbf{u}_e(\mathbf{x})) = \mathbf{0}$ so that

$$\delta\dot{\mathbf{x}} = \mathbf{E}\,\delta\mathbf{x} + \mathbf{D}\,\delta\mathbf{u}, \tag{4.10}$$

where

$$\mathbf{E} = \frac{\partial \mathbf{f}}{\partial \mathbf{x}}(\mathbf{x}, \mathbf{u}_e(\mathbf{x}))$$

and

$$\mathbf{D} = \frac{\partial \mathbf{f}}{\partial \mathbf{u}}(\mathbf{x}, \mathbf{u}_e(\mathbf{x})).$$

Formally, we can express *local controllability* of nonlinear system (4.9) in terms of controllability of (4.10). Hence the argument suggests that the linear system rank test (3.93) in Vol. 1 be applied to (4.10). That is, if the rank of the following matrix is n, the order of the system, then (4.9) is locally controllable:

$$\mathbf{M}_\delta = [\mathbf{D} \quad \mathbf{ED} \quad \cdots \quad \mathbf{E}^{n-1}\mathbf{D}]. \tag{4.11}$$

Or more meaningful perhaps, (4.9) is *locally controllable* on the equilibrium set if there exists a $\mathbf{v}$ such that $\mathbf{Dv}$ lies in no invariant subspace of dimension at most $(n - 1)$ of the matrix $\mathbf{E}$.

This result is put on a more rigorous basis in [L5]. The useful result for application follows.

Theorem 2. The reachable zone $R(\mathbf{x})$ and the incident zone $I(\mathbf{x})$ from the equilibrium set given by $\mathbf{f}(\mathbf{x}, \mathbf{u}_e(\mathbf{x})) = \mathbf{0}$, where $\mathbf{u}_e(\mathbf{x})$ is interior to U, are open connected subsets of $\mathbb{R}^n$ if there exists a $\mathbf{v} \in \mathbb{R}^m$ such that $\mathbf{Dv}$ lies in no invariant subspace of $\mathbf{E}$ of dimension at most $(n - 1)$.

See [L5] for the proof. If the equilibrium set is a connected proper subset of $\mathbb{R}^n$ and Theorem 2 is satisfied for every interior $\mathbf{u} \in U$, the reachable zone from any equilibrium state $\mathbf{x}$ corresponding to an interior control value $\mathbf{u}_e(\mathbf{x})$ includes every other equilibrium state corresponding to an interior control value of U. For if $R(\mathbf{x})$ has an equilibrium state $\mathbf{y}$ on its boundary, $I(\mathbf{y})$ must be disjoint from $R(\mathbf{x})$. But by Theorem 2, $I(\mathbf{y})$ is an open neighborhood of $\mathbf{y}$ if $\mathbf{u}_e(\mathbf{y})$ is in the interior of U. Hence $\mathbf{u}_e(\mathbf{y})$ must be on the boundary of U.

For the general bilinear system, the matrix $\mathbf{D}$ is

$$\mathbf{D} = \frac{\partial \mathbf{f}}{\partial \mathbf{u}}(\mathbf{x}, \mathbf{u}_e(\mathbf{x})) = [\mathbf{N}_1\mathbf{x} \mid \mathbf{N}_2\mathbf{x} \mid \cdots \mid \mathbf{N}_m\mathbf{x}] + \mathbf{B} \tag{4.12}$$

and $\mathbf{E}$ is defined by

$$\mathbf{E} = \frac{\partial \mathbf{f}}{\partial \mathbf{x}}(\mathbf{x}, \mathbf{u}_e(\mathbf{x})) = \mathbf{A} + \sum_{k=1}^{m} u_k(\mathbf{x})\mathbf{N}_k. \tag{4.13}$$

Substitution of the expression for the equilibrium state into (4.13) yields the theorem for a bilinear system: The system (3.1) is locally controllable at the equilibrium state corresponding to an interior $\mathbf{u} \in U$ if there exists a $\mathbf{v} \in \mathbb{R}^m$ such that

$$-\sum_{j=1}^{m} \mathbf{v}_j \left[\mathbf{N}_j(\mathbf{A} + \sum_{k=1}^{m} u_k \mathbf{N}_k)^{-1} \mathbf{B}\mathbf{u} \right] + \mathbf{B}\mathbf{v}$$

lies in no invariant subspace of dimension at most $(n - 1)$ of the matrix $\mathbf{E}$.

The practical implications of using this criterion for a general bilinear system appear quite formidable. However, it will now be shown that the application to systems of the phase-variable type is straightforward, and in fact such systems always satisfy the criterion when $\mathbf{B} \neq \mathbf{0}$.

The maxtrix $\mathbf{D}$ for the phase-variable system is obtained by substituting the equilibrium-state expression into (4.12). Hence it can be shown that

$$\mathbf{D} = \frac{-\sum_{k=1}^{m} b_k u_k}{a_1 + \sum_{k=1}^{m} u_k n_{1k}} \begin{bmatrix} 0 & 0 & \cdots & 0 \\ 0 & 0 & & \vdots \\ \vdots & \vdots & & \vdots \\ n_{11} & n_{12} & \cdots & n_{1m} \end{bmatrix} + \mathbf{B}. \tag{4.14}$$

The matrix $\mathbf{E}$ is simply of the canonical phase-variable form.

The criterion to be satisfied is that a $\mathbf{v} \in \mathbb{R}^m$ is to be found such that the vectors $\mathbf{Dv}, \mathbf{EDv}, \ldots, \mathbf{E}^{(n-1)}\mathbf{Dv}$ are linearly independent. By inspection of (4.14), if $\mathbf{D}$ is not identically zero, it has nonzero entries only in the bottom row, and there exists a $\mathbf{v}$ such that $\mathbf{Dv}$ has the nth component nonzero and all others zero, $\mathbf{E}^2\mathbf{Dv}$ has the $(n - 2)$th component nonzero, and so on. Such a set is certainly linearly independent.

There remains the question of whether $\mathbf{D}$ is identically zero. By inspection of (4.14), if $\mathbf{B} = \mathbf{0}$, then $\mathbf{D} = \mathbf{0}$ and the criterion is not satisfied. Suppose that $\mathbf{B} \neq \mathbf{0}$, and for the particular value of $\mathbf{u}$ under consideration,

$$\sum_{k=1}^{m} b_k u_k = 0.$$

Then $\mathbf{D} = \mathbf{B}$ and the criterion is satisfied. Finally, if

$$\sum_{k=1}^{m} b_k u_k \neq 0,$$

select $\mathbf{v} = \mathbf{u}$ and then

$$\mathbf{Dv} = \mathbf{Du} = \begin{bmatrix} 0 \\ 0 \\ \vdots \\ \sum_{k=1}^{m} b_k u_k \left(1 - \dfrac{-\sum_{k=1}^{m} n_{1k} u_k}{a_1 + \sum_{k=1}^{m} n_{1k} u_k}\right) \end{bmatrix}.$$

This is not identically zero since **A** nonsingular implies that $a_1 \neq 0$. Therefore, the criterion is always satisfied if **B** is not identically zero for phase-variable systems.

Now if it can be shown that the equilibrium set is connected and the system eigenvalues of (3.1), as **u** ranges over U, can all be driven from the left-half plane to the right-half plane (or vice versa), the BLS (3.1) should be controllable. To verify this further, assume a fixed control value $\mathbf{u}^-$ in the interior of U such that the eigenvalues of the system matrix

$$\mathbf{A} + \sum_{k=1}^{m} u_k^- \mathbf{N}_k$$

all have negative real parts. Then the trajectories of the system with constant control $\mathbf{u} = \mathbf{u}^-$ cover all the $\mathbb{R}^n$, and each trajectory approaches the unique equilibrium state

$$\mathbf{x}_e(\mathbf{u}^-) = -\left(\mathbf{A} + \sum_{k=1}^{m} u_k^- \mathbf{N}_k\right)^{-1} \mathbf{B}\mathbf{u}^-. \tag{4.15}$$

For any initial state $\mathbf{x}_0 \in \mathbb{R}^n$, the unique trajectory passing through $\mathbf{x}_0$ with control $\mathbf{u}^-$ reaches any neighborhood of $\mathbf{x}_e(\mathbf{u}^-)$ in finite time.

Suppose, also, that there exists a fixed control value $\mathbf{u}^+$ in the interior of U such that the eigenvalues of the matrix

$$\mathbf{A} + \sum_{k=1}^{m} u_k^+ \mathbf{N}_k$$

all have positive real parts. Then the trajectories of the system with constant control $\mathbf{u} = \mathbf{u}^+$ cover all of the $\mathbb{R}^n$, and each trajectory corresponds to motion away from the unique equilibrium state

$$\mathbf{x}_e(\mathbf{u}^+) = -\left(\mathbf{A} + \sum_{k=1}^{m} u_k^+ \mathbf{N}_k\right)^{-1} \mathbf{B}\mathbf{u}^+. \tag{4.16}$$

For any terminal state $\mathbf{x}_f$, the unique trajectory passing through $\mathbf{x}_f$ with control $\mathbf{u}^+$ reaches $\mathbf{x}_f$ from any neighborhood of $\mathbf{x}_e(\mathbf{u}^+)$ in finite time.

If $\mathbf{u}^-$ and $\mathbf{u}^+$ both exist, and if every point of some neighborhood of $\mathbf{x}_e(\mathbf{u}^+)$ can be reached from every point of some neighborhood of $\mathbf{x}_e(\mathbf{u}^-)$, then certainly any terminal state $\mathbf{x}_f$ can be reached from any initial state $\mathbf{x}_0$, and the system is certainly controllable. Such equilibrium sets and their connectedness are described in [R1] and [M2]. Then the controllability analysis can be completed by means of the local controllability Theorem 2, which provides sufficient conditions for local controllability of a system in a neighborhood of its equilibrium set.

The controllability analysis presented here can be summarized by the following sufficient conditions.

Theorem 3. The bilinear system (3.1) is completely controllable if:

(a) there exist control values $\mathbf{u}^+ \in U$ and $\mathbf{u}^- \in U$ such that the real parts of the eigenvalues of the system matrix are positive and negative, respectively, and such that equilibrium states $\mathbf{x}_e(\mathbf{u}^+)$ and $\mathbf{x}_e(\mathbf{u}^-)$ are contained in a connected component of the equilibrium set;

(b) for each $\mathbf{x}$ in the equilibrium set with an equilibrium control $\mathbf{u}_e(\mathbf{x}) \in U$ such that if $\mathbf{f}(\mathbf{x}, \mathbf{u}_e(\mathbf{x})) = \mathbf{0}$, there exists a $\mathbf{v} \in \mathbb{R}^m$ such that $\mathbf{g}$ lies in no invariant subspace of dimension at most $(n - 1)$ of the matrix $\mathbf{E}$, where $\mathbf{g} = \mathbf{D}\mathbf{v}$; $\mathbf{D}$ is given by (4.12) and $\mathbf{E}$ by (4.13).

See [R1] and [M2] for details.

Remark 1. For phase-variable systems, $x_1 = x$, $x_2 = \dot{x}$, . . . , $x_n = x^{(n-1)}$, condition (b) is always satisfied if $\mathbf{B}$ is a nonzero matrix.

Remark 2. Condition (a) is satisfied if all the eigenvalues of the system matrix

$$\mathbf{A} + \sum_{k=1}^{m} u_k \mathbf{N}_k$$

can be shifted across the imaginary axis of the complex plane without passing through zero, as $\mathbf{u}$ ranges continuously over a subset of U.

It can be shown that for phase-variable systems, condition (a) of the main result may be satisfied even if one or more eigenvalues pass through zero. On the other hand, this condition cannot be satisfied with scalar control ($m = 1$) in a state space of odd dimension n. For then an odd number of eigenvalues must be shifted across the imaginary axis, and at least one of these must pass through zero, since at most

$(n - 1)/2$ can cross as complex-conjugate pairs. However, it can be shown that for $m = 1$ that the branches of the equilibrium curve are disjoint.

Example 4.4

Consider the second-order system

$$\dot{x}_1 = x_2,$$
$$\dot{x}_2 = -x_1 - x_2 + u,$$

which is completely controllable for unbounded control. For bounded control, however, this system is locally controllable only in some vicinity of the origin. If the process admits an appropriate bilinear mode of control, the system is completely controllable since the local stability criterion Theorem 2 is already satisfied. For example, consider the bilinear system

$$\dot{x}_1 = x_2,$$
$$\dot{x}_2 = -x_1 - (1 + 4u)x_2 + u,$$

where $|u| \leq 1$ and the eigenvalues

$$\lambda_{1,2}(u) = \frac{-(1 + 4u)}{2} \pm \frac{[(1 + 4u)^2 - 4]^{1/2}}{2}$$

for constant u. For $u = -1$ both are real and positive, crossing the imaginary axis as a complex-conjugate pair. Thus u^+ and u^- exist and the equilibrium set is connected. The system is of the phase-variable type and $\mathbf{B} \neq \mathbf{0}$. Thus the new system is completely controllable. For example, with $u^+ = -\frac{1}{2}$ and $u^- = 0$.

As shown by the next example, however, not all bilinear systems with the capability of transferring system eigenvalues across the imaginary axis are completely controllable. Such is the case even though the system satisfies the local-controllability criterion.

Example 4.5 [M2]

Consider

$$\dot{x}_1 = 2ux_1 + x_2,$$
$$\dot{x}_2 = x_1 + 2ux_2 + u,$$

with eigenvalues $\lambda_1(u) = 2u + 1$, $\lambda_2(u) = 2u - 1$. If the constraint set U is given by $|u| \leq 1$, the values u^+ and u^- exist. It is easily verified that Theorem 2 is satisfied for this system. The equilibrium set, however, is not connected, since $m = 1$, and both eigenvalues pass through zero as they cross the imaginary axis in the complex plane. Therefore, the criteria for complete controllability are not all satisfied.

The phase-plane portraits for $u = +1$ and $u = -1$ are shown superimposed in Fig. 4.2. The solid trajectories are for the system with $u = -1$, which has a stable node at $x_1 = -\frac{1}{3}$, $x_2 = -\frac{2}{3}$. The dashed trajectories are for the system with $u = +1$, which has an unstable node at $x_1 = -\frac{1}{3}$, $x_2 = -\frac{2}{3}$. The equilibrium set is shown as the three heavy solid curves as u ranges from $u = -1$ to $u = +1$. The reader should compute this equilibrium set as an exercise.

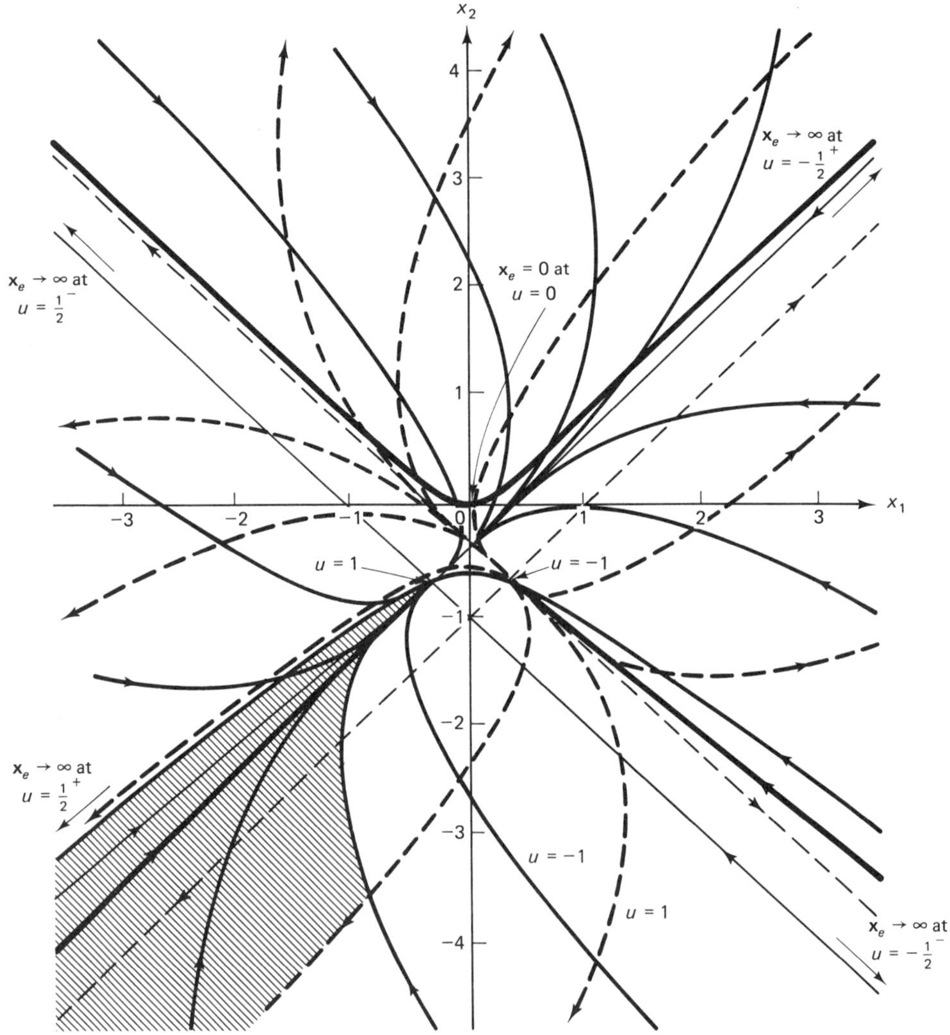

Figure 4.2 Bilinear phase trajectories for Example 4.5.

The reachable zones from various initial phases are easily determined on this figure simply by considering the directions of allowable motion at each point, which is the cone between the extremal directions corresponding to $u = +1$ and $u = -1$. It can be shown that the system is not completely controllable. For example, the reachable zone from the point y is just the shaded region with $u = \pm 1$.

A bilinear system is completely controllable if the conditions given by the main result (Theorem 3) are satisfied. As one might expect, the conditions are not as simple as the popular conditions for complete controllability of linear systems with unconstrained control or for null controllability with constrained control. For phase-variable systems, however, the sufficient conditions are easy to apply.

In practice, a bilinear mode may be implemented by controlling significant plant parameters in a manner similar to the variable wing geometry of high-performance aircraft. For Example 4.3, a simple bilinear control made a locally controllable linear system completely controllable. With respect to many systems that are inherently bilinear, however, controllability is further complicated by numerous state constraints. Unfortunately, state constraints can appear in such diversity that meaningful controllability conditions would have to be specified for particular cases. For the classical neutron kinetics of a reactor, it is obvious by inspection of the model that every desired equilibrium state can be attained in finite time. Again such controllability is a consequence of the bilinear control mode, and is analyzed along with optimal control in Chapter 6. Similar to the nuclear reactor, the human immune system synthesizes controlled use of instability to attain sufficient controllability to overcome replicating viruses. For example, T-helper cells turn-on a divergent mode which is turned off by suppressor cells.

The inherent relationship between controllability and stability for nonlinear systems which are linear in control is studied in [G4]. This work is based on the basic concept of controllability and the use of Liapunov functions similar to that above. In this manner, sufficiency conditions for complete controllability and for uncontrollability based on the existence of Liapunov-like functions are derived.

4.2.3 Other Forms of Controllability

As for linear systems, the bilinear system (3.1) may be decomposed into subsystems that are controllable and uncontrollable. Similarly with an output, say $\mathbf{y} = \mathbf{C}\mathbf{x}$, the system may be further decomposed into subsystems that are (1) controllable and observable, (2) controllable but not observable, (3) observable but not controllable, and (4) not controllable and not observable. Again, if we are synthesizing a state-space realization from input–output data, we are normally interested in a minimum realization. The system is a minimum realization if and only if the system is controllable and observable. Here, of course, we are assuming that the control can have infinite magnitude, that is, controllability in the traditional sense with all of state space attainable without the bounded admissible control clause.

Most physical BLS have independent additive and multiplicative control in which case (3.1) can be written as

$$\frac{d\mathbf{x}}{dt} = \mathbf{A}\mathbf{x} + \sum_{k=1}^{m-q} \mathbf{N}_k u_k \mathbf{x} + \mathbf{B}\mathbf{v}, \tag{4.17}$$

where $\mathbf{v} \in \mathbb{R}^q$, and $\mathbf{A}$, $\mathbf{B}$, $\mathbf{N}_k$, $k = 1, \ldots, m - q$, are constant matrices of appropriate dimension. Suppose, for example, that $n = 2$, $q = 1$, $m = 2$, $\mathbf{A} = \mathbf{0}$,

$$\mathbf{B} = \mathbf{b} = \begin{bmatrix} 1 \\ 1 \end{bmatrix}, \qquad \mathbf{N}_1 = \mathbf{N} = \begin{bmatrix} 1 & -1 \\ 0 & 0 \end{bmatrix}.$$

Then the transformation $\mathbf{x} = \mathbf{T}\mathbf{y}$ with

$$\mathbf{T} = \begin{bmatrix} 1 & 1 \\ 1 & 0 \end{bmatrix}$$

yields

$$\frac{d\mathbf{y}}{dt} = \begin{bmatrix} 1 \\ 0 \end{bmatrix}\mathbf{v} + \begin{bmatrix} 0 & 0 \\ 0 & 1 \end{bmatrix}\mathbf{y}u.$$

Obviously, the system is not controllable. The $y_2 = 0$ axis is the subspace that can be reached from any exterior state, but no control will cause a trajectory to leave this noncontrollable subspace.

If (4.17) has unconstrained additive scalar control and magnitude-constrained multiplicative scalar control, it can be seen from linear-system controllability that (4.17) is not controllable if and only if the pairs ($\mathbf{A}$, $\mathbf{B}$) and ($\mathbf{B}$, $\mathbf{N}$) are not controllable. Then it is sufficient that either ($\mathbf{A}$, $\mathbf{B}$) or ($\mathbf{B}$, $\mathbf{N}$) be controllable for (4.17) to be controllable. In general, it is not necessary that both be controllable for the BLS to be controllable.

References [K3] and [K7] study controllability of a special case of (4.17) and of (3.1). Here Lie algebra [C5] is used to define the attainable states in terms of an integral manifold which is reachable by piecewise-constant controls in the presence of control magnitude constraints.

In [H1], Pfaffian equations [C1] are shown to be naturally associated with BLS and more general nonlinear systems that are linear in control. Then controllability is derived from integrability of the associated Pfaffian equations. Unfortunately, all of these results are difficult to apply and beyond the scope of this book.

Lie algebra has been applied successfully to the study of controllability of BLS and other nonlinear systems ([K3], [K7], [B3], [H8], [S4], and [H4]). Again, most of these results are difficult to apply and beyond the scope of the present treatment.

A few basic concepts, however, are summarized in a formal manner below. The intent here is to summarize the results without unnecessary complications. Many readers may wish to bypass the remainder of this section since it is not used elsewhere in the book. Those with a desire to understand the differential-geometrical approach more thoroughly are referred to the excellent references.

First, let us consider several alternative definitions of controllability and other useful terms in order to make this theory very useful [H4]. Again consider the general nonlinear system (4.9) with $\mathbf{u} \in U$ a subset of $\mathbb{R}^m$ with $\mathbf{f}(\cdot)$ analytic (all derivatives exist).

Accessibility. Given an open subset X of $\mathbb{R}^n$, $\mathbf{x}' A_x \mathbf{x}^0$ (i.e., $\mathbf{x}'$ is x-accessible from $\mathbf{x}^0$) if there exists a $\mathbf{u}(t) \in U$ which steers $\mathbf{x}(t)$ from $\mathbf{x}^0$ to $\mathbf{x}'$ in finite time, where $\mathbf{x}(t) \in A_x$ satisfies (4.9). If A_x includes all of $\mathbb{R}^n$, we say that $\mathbf{x}$ is accessible. The set of all points accessible from $\mathbf{x}^0$ is written as $A(\mathbf{x}^0)$. Then we say that (4.9) is *controllable at* $\mathbf{x}^0$ if $A(\mathbf{x}^0) \equiv \mathbb{R}^n$, and controllable if $A(\mathbf{x}) \equiv \mathbb{R}^n$ for every $\mathbf{x} \in \mathbb{R}^n$. Here it is assumed that acceptable solutions to (4.9) are found for all state space, $\mathbb{R}^n$.

For (4.9) to be controllable may require that $\mathbf{x}(t)$ take on prohibitively large variations in order to realize its controllability mission. Consequently, it is convenient to define a new concept of local controllability [H5] which is sometimes called local-local controllability [H4]. Here we refer to this form of local controllability as strong local controllability. We say that (4.9) is *strongly locally controllable* at $\mathbf{x}^0$ if for every neighborhood $\mathbf{x}$ of $\mathbf{x}^0$, $A_x(\mathbf{x}^0)$ is also a neighborhood of $\mathbf{x}^0$. The system (4.9) is *strongly locally controllable* if it is strongly locally controllable at every $\mathbf{x} \in \mathbb{R}^n$.

Weak accessibility. Given a sequence of states $\mathbf{x}^0, \ldots, \mathbf{x}^k \in \mathrm{X}$ such that $\mathbf{x}^i A_x \mathbf{x}^{i-1}$ or $\mathbf{x}^{i-1} A_x \mathbf{x}^i$, for $i = 1, \ldots, k$, then $\mathbf{x}^0\, WA_x \mathbf{x}^k$. That is, $\mathbf{x}^k$ is *weakly* $\mathbf{x}$-*accessible on* $\mathbf{X}$ if $\mathbf{x}^i$ is accessible from $\mathbf{x}^{i-1}$ or $\mathbf{x}^{i-1}$ is accessible from $\mathbf{x}^i$ on X for all $i = 1, \ldots, k$. $WA(\mathbf{x}^0)$ denotes weak accessibility at $\mathbf{x}^0$, and (4.9) is *weakly controllable* at $\mathbf{x}^0$ if $WA(\mathbf{x}^0) = \mathbb{R}^n$, which in this case is identical to $WA(\mathbf{x}) = \mathbb{R}^n$. Furthermore, (4.9) is said to be *locally weakly controllable at* $\mathbf{x}^0$ if for every neighborhood $\mathbf{X}$ of $\mathbf{x}^0$, $WA_x(\mathbf{x}^0)$ is a neighborhood of $\mathbf{x}^0$. Then (4.9) is *locally weakly controllable* if it is locally weakly controllable at all $\mathbf{x} \in \mathbb{R}^n$. In general, then, strong local controllability is the strongest requirement with weak controllability the least requirement. Controllability is perhaps more significant than local weak controllability, but the two involve different concepts. Their relative implications follow:

$$
\begin{array}{ccc}
\text{strong local controllability} & \Rightarrow & \text{controllability} \\
\Downarrow & & \Downarrow \\
\text{local weak controllability} & \Rightarrow & \text{weak controllability}
\end{array}
$$

For autonomous linear systems these controllability definitions are identical. Note that in Section 4.2.2, local controllability is, of course, weaker than strong local controllability or controllability. However, it is a different concept than local weak controllability or weak controllability. It will include the latter two, however, if there is local controllability for all $\mathbf{x} \in \mathbb{R}^n$ rather than just the equilibrium set.

Here we follow the nice interpretation of the results of [H5], which is given in [V3]. The computational procedure is as follows:

1. Let $F_0 = \mathbf{f}(\mathbf{x}, \mathbf{u})$ for constant $\mathbf{u} \in \mathbb{R}^m$.
2. Compute all the possible Lie brackets $[f_1, f_2] \equiv f_1 f_2 - f_2 f_1$ for all $f_1, f_2 \in F_0$. Designate $\bar{F}_1 \equiv$ the set of all these vector fields.
3. Let $F_1 = F_0 \cup \bar{F}_1$.
4. Compute all possible $[f_1, f_2]$ for all $f_1, f_2 \in \bar{F}_1$. This yields the totality of these vector fields denoted by $\bar{F}_2$.
5. Repeat steps 1 to 4 until $F(\cdot) \equiv$ the linear combination of all vector fields for (4.9). For the BLS this is always possible with a finite number of steps. $F(\cdot)$ equivalently is defined as the Lie algebra of $\mathbf{f}(\mathbf{x}, \mathbf{u})$ as $\mathbf{u}$ ranges over $\mathbb{R}^m$.

Here the algebra is simplified if (3.1) is assumed to be homogeneous. That is,

$$\frac{d\mathbf{x}}{dt} = \left(\mathbf{A} + \sum_{k=1}^{m} u_k \mathbf{N}_k\right)\mathbf{x}. \tag{4.18}$$

Then each fixed value of $\mathbf{u}$ yields a summation of $(m + 1)$ linear vector fields $\mathbf{Ax}$, $u_1\mathbf{N}_1\mathbf{x}, \ldots, u_m\mathbf{N}_m\mathbf{x}$. For the foregoing procedure, the following steps result:

1. $F_0 \equiv$ the span of $(m + 1)$ vector fields $\mathbf{Ax}, \mathbf{N}_1\mathbf{x}, \ldots, \mathbf{N}_m\mathbf{x}$.
2. Then the Lie bracket of the vector fields is given by $[\mathbf{Ax}, \mathbf{N}_k\mathbf{x}] = [\mathbf{A}, \mathbf{N}_k]\mathbf{x} = (\mathbf{AN}_k - \mathbf{N}_k\mathbf{A})\mathbf{x}$, for $k = 1, \ldots, m$. This yields $\bar{F}$.
3. $F_1 \equiv F_0 \cup \bar{F} = \mathbf{Ax}, \mathbf{Nx}, \ldots, \mathbf{N}_m\mathbf{x}, (\mathbf{AN}_1 - \mathbf{N}_1\mathbf{A})\mathbf{x}, \ldots, (\mathbf{AN}_m - \mathbf{N}_m\mathbf{A})\mathbf{x}$.
4. $F_2 \equiv [f_1, f_2]$ for all $f_1, f_2 \in F_1$, and so on.

Then n^2 linearly independent matrices would imply weak controllability of (4.18).

Example 4.6 [V3]

Consider the special homogeneous BLS

$$\dot{\mathbf{x}} = (\mathbf{A} + \mathbf{I}u)\mathbf{x},$$

where u is a scalar control. It is readily seen to be not controllable with magnitude-constrained u by the methods of Section 4.2.2. To check weak controllability, let $F_0 = (\mathbf{Ax}, \mathbf{x})$, so that F_1 yields $[\mathbf{A}, \mathbf{I}] = \mathbf{A} - \mathbf{A} = \mathbf{0}$. The process is aborted since $\mathbf{0}$ is not linearly independent of $(\mathbf{A}, \mathbf{I})$. Therefore, no n^2 linearly independent matrices can be found from the procedure.

Example 4.7 [V3]

Let

$$\mathbf{A} = \begin{bmatrix} 0 & 1 \\ -1 & 0 \end{bmatrix}, \qquad \mathbf{N} = \mathbf{N}_1 = \begin{bmatrix} 1 & 0 \\ 0 & 2 \end{bmatrix}$$

with $n = 2$, $m = 1$ in (4.18). Now

a. $F_0 = [\mathbf{Ax}, \mathbf{Nx}] = \left\{ \begin{bmatrix} 0 & 1 \\ -1 & 0 \end{bmatrix} \mathbf{x}, \begin{bmatrix} 1 & 0 \\ 0 & 2 \end{bmatrix} \mathbf{x} \right\}.$

b. $\bar{F} : \mathbf{C} = [\mathbf{A}, \mathbf{N}] = \mathbf{AN} - \mathbf{NA} = \begin{bmatrix} 0 & 1 \\ 1 & 0 \end{bmatrix}.$

c. $F_1 \equiv F_0 \cup \bar{F}.$

d. $F_2 \equiv$ all possible combinations of $\{\mathbf{A}, \mathbf{N}, \mathbf{C}\}$,

$$\mathbf{D} = [\mathbf{A}, \mathbf{C}] = \begin{bmatrix} 3 & 0 \\ 0 & -3 \end{bmatrix}, \qquad [\mathbf{N}, \mathbf{C}] = -\mathbf{A}.$$

Then it is seen that **A**, **N**, **C**, and **D** are linearly independent matrices such that $n^2 = 4$ and the system is weakly controllable. Again, however, the system is not controllable in the normal sense. This example shows the shortcoming of weak controllability. It is a relatively simple matter to evaluate the solution $\mathbf{x}(t)$ for various constant u. For example, $u = 0$ yields an oscillator (center portrait); $u = \pm 1$, a stable and unstable focus, and as u becomes large, a saddle-point follows. With $u = \pm 1$, the system is controllable on the zero-punctured state plane, $\{\mathbb{R}^2 - 0\}$. That is, the system is controllable relative to every state except the origin, which is an equilibrium state.

Although not of major interest here, *weak observability* is equivalent to every initial state being distinguishable from every nearby initial state. Here $\mathbf{x}^0$ and $\mathbf{x}^1$ are distinguishable if there exists some $\mathbf{u}(t)$ such that the output trajectories $\mathbf{y} = \mathbf{g}(\mathbf{x})$ for (4.9) which start at the two different initial states $\mathbf{x}^0$ and $\mathbf{x}^1$ are different. Then, in general, the system is *observable* if every possible pair of initial states is distinguishable. A sufficiency rank test can be made for weak observability, which again depends on appropriate linearly independent vectors defined by a Lie algebra ([H5] and [V3]).

Another approach to observability for nonlinear systems depends on the inverse function theorem which is associated with $\mathbf{x} = \mathbf{g}^{-1}(\mathbf{y})$ [P3]. Here a necessary condition is to evaluate the rank of the corresponding Jacobian, which if of value n also provides a form of local observability ([K8] and [K9]). In [M20] a global version of nonlinear observability is introduced which uses the inverse function theorem. Unfortunately, the method generally is quite cumbersome and not developed rigorously at this time.

4.3 OBSERVER CONSTRUCTION

Linear system observers were introduced in Vol. 1, Section 3.6. Such Luenberger observers are designed after [L6]. In the present section, a reduced-order BLS observer is formulated in a manner that follows [H6] and [D1]. Other formulations of BLS observers are presented in [D1], [W1], and [H7].

First consider the linear time-varying system of the form

$$\dot{\mathbf{x}} = \mathbf{A}(t)\mathbf{x} + \mathbf{B}(t)\mathbf{u}, \qquad \mathbf{y} = \mathbf{C}(t)\mathbf{x}, \tag{4.19}$$

where $\mathbf{x} \in \mathbb{R}^n$, $\mathbf{u} \in \mathbb{R}^m$, and $\mathbf{y} \in \mathbb{R}^p$. Assume that the system is controllable and observable.

Practically, it is desired to reconstruct a linear function of the state $\mathbf{Kx}(t)$; typically, a linear feedback control law of the form

$$\mathbf{u} = \mathbf{Kx} \tag{4.20}$$

is used, where $\mathbf{K}$ is an $m \times n$ matrix.

To achieve this state-function reconstruction, an observer of the form

$$\begin{aligned} \dot{\mathbf{z}} &= \mathbf{D}(t)\mathbf{z} + \mathbf{E}(t)\mathbf{y} + \mathbf{G}(t)\mathbf{u}, \\ \mathbf{w} &= \mathbf{P}(t)\mathbf{z} + \mathbf{V}(t)\mathbf{y} \end{aligned} \tag{4.21}$$

will be designed according to the Luenberger observer [L6], where $\mathbf{z} \in \mathbb{R}^r$ and $\mathbf{w} \in \mathbb{R}^q$ are used to approximate $\mathbf{Kx}$. Since (4.21) is a reduced-order observer, given p linear state combinations in the output (4.19), we wish to generate the remaining $n - p$ linear state combinations

$$\mathbf{z} = \mathbf{T}(t)\mathbf{x}, \qquad \mathbf{z}(t) \in \mathbb{R}^{n-p}. \tag{4.22}$$

Combining (4.19) and (4.22) gives us

$$\begin{bmatrix} \mathbf{y} \\ \mathbf{z} \end{bmatrix} = \begin{bmatrix} \mathbf{C}(t) \\ \mathbf{T}(t) \end{bmatrix} \mathbf{x}. \tag{4.23}$$

Then the complete state vector is introduced exactly through

$$\mathbf{Kx} = \begin{bmatrix} \mathbf{C}(t) \\ \mathbf{T}(t) \end{bmatrix}^{-1} \begin{bmatrix} \mathbf{y} \\ \mathbf{z} \end{bmatrix} = [\mathbf{V}(t) \quad \mathbf{P}(t)] \begin{bmatrix} \mathbf{y} \\ \mathbf{z} \end{bmatrix}. \tag{4.24}$$

Therefore, the output $\mathbf{w}(t)$ of (4.21) is said to estimate $\mathbf{Kx}(t)$ in an asymptotic sense if

$$\lim_{t \to \infty} [\mathbf{w}(t) - \mathbf{K}\mathbf{x}(t)] = \mathbf{0}. \tag{4.25}$$

Since $\mathbf{z}(t)$ is acceptable as an approximation of $\mathbf{T}(t)\mathbf{x}(t)$, the observer reconstruction error can be characterized by

$$\boldsymbol{\epsilon} = \mathbf{z} - \mathbf{T}(t)\mathbf{x}. \tag{4.26}$$

Using (4.19), (4.21), and (4.26) the dynamics of the observer error is

$$\begin{aligned} \dot{\boldsymbol{\epsilon}} &= \dot{\mathbf{z}} - \dot{\mathbf{T}}(t)\mathbf{x} - \mathbf{T}(t)\dot{\mathbf{x}} \\ \dot{\boldsymbol{\epsilon}} &= \mathbf{D}(t)\boldsymbol{\epsilon} + [\mathbf{D}(\mathbf{t})\mathbf{T}(t) - \mathbf{T}(t)\mathbf{A}(t) + \mathbf{E}(t)\mathbf{C}(t) - \dot{\mathbf{T}}(t)]\mathbf{x} \\ &\quad + [\mathbf{G}(t) - \mathbf{T}(t)\mathbf{B}(t)]\mathbf{u}. \end{aligned} \tag{4.27}$$

Equation (4.27) can be reduced to the homogeneous equation

$$\dot{\boldsymbol{\epsilon}} = \mathbf{D}(t)\boldsymbol{\epsilon}(t) \tag{4.28}$$

provided that the observer matrices $\mathbf{T}(t)$, $\mathbf{D}(t)$, $\mathbf{E}(t)$, and $\mathbf{G}(t)$ satisfy

$$\mathbf{D}(t)\mathbf{T}(t) - \mathbf{T}(t)\mathbf{A}(t) = \dot{\mathbf{T}}(t) - \mathbf{E}(t)\mathbf{C}(t) \tag{4.29}$$

and

$$\mathbf{G}(t) = \mathbf{T}(t)\mathbf{B}(t). \tag{4.30}$$

Also, from (4.24) we have that

$$\mathbf{K}\begin{bmatrix} \mathbf{C}(t) \\ \mathbf{T}(t) \end{bmatrix}^{-1} = [\mathbf{V}(t) \quad \mathbf{P}(t)]. \tag{4.31}$$

Equations (4.28)–(4.31) are known as the observer constraint equations which the parameter matrices $\mathbf{T}(t)$, $\mathbf{D}(t)$, $\mathbf{E}(t)$, $\mathbf{P}(t)$, and $\mathbf{V}(t)$ of the reduced-order observer (4.21) are required to satisfy.

Furthermore, denote the actual state reconstruction error by

$$\mathbf{e}(t) \equiv \hat{\mathbf{x}}(t) - \mathbf{x}(t) \tag{4.32}$$

which reduces to

$$\mathbf{e}(t) = \mathbf{P}(t)\boldsymbol{\epsilon}(t). \tag{4.33}$$

Thus the accuracy of the state reconstruction is a linear function of that of the observer state vector $\mathbf{z}(t)$. Then from (4.28)–(4.31), it can be seen that the q-dimensional observer (4.29)–(4.30) will estimate $\mathbf{Kx}(t)$ in the sense of (4.25), as summarized by the following:

1. $\mathbf{D}(t)$ is a stable matrix.
2. $\mathbf{D}(t)\mathbf{T}(t) - \mathbf{T}(t)\mathbf{A}(t) = \dot{\mathbf{T}}(t) - \mathbf{E}(t)\mathbf{C}(t)$.
3. $\mathbf{G}(t) = \mathbf{T}(t)\mathbf{B}(t)$.
4. $\mathbf{K}(t) = \mathbf{P}(t)\mathbf{T}(t) + \mathbf{V}(t)\mathbf{C}(t)$. (4.34)

Next consider an n-dimensional controllable and observable bilinear system represented by (3.1) or (4.8) and output

$$\mathbf{y} = \mathbf{Cx}, \tag{4.35}$$

where $\mathbf{y} \in \mathbb{R}^p$, $\mathbf{C}$ is a $p \times n$ matrix, and rank $\mathbf{C} = p$. Since we can consider (4.8) as a time-varying linear system here, and from (4.29) matrices $\mathbf{D}(t)$ and $\mathbf{E}(t)$ are related to matrices $\mathbf{A}(t)$, we can construct the following observer, which includes $u_k\mathbf{z}$ and $u_k\mathbf{y}$ terms.

$$\dot{\mathbf{z}} = \mathbf{D}^0\mathbf{z} + \sum_{k=1}^{m} \mathbf{D}^k u_k \mathbf{z} + \mathbf{E}^0\mathbf{y} + \sum_{k=1}^{m} \mathbf{E}^k u_k \mathbf{y} + \mathbf{Gu}, \tag{4.36}$$

$$\mathbf{w} = \mathbf{Pz} + \mathbf{Vy}, \tag{4.37}$$

where $(\mathbf{D}^0, \mathbf{P})$ is an observable pair, $\mathbf{z} \in \mathbb{R}^r$ and $\mathbf{w} \in \mathbb{R}^q$, $\mathbf{D}^k$, $\mathbf{E}^k$, $k = 0, 1, \ldots, m$, $\mathbf{G}$, $\mathbf{P}$, and $\mathbf{V}$ are of appropriate dimensions. The relationships (4.20)–(4.24) still hold except that the matrices $\mathbf{C}$, $\mathbf{T}$, $\mathbf{V}$, and $\mathbf{P}$ are constant.

Since the observer for the bilinear system with estimation error independent of input and initial states is considered, we can construct a bilinear observer based on the following definition: (4.36)–(4.37) is said to be a q-dimensional observer of $\mathbf{Kx}$ of (4.8) and (4.35) if and only if

$$\lim_{t \to \infty} [\mathbf{w}(t) - \mathbf{Kx}(t)] = \mathbf{0}$$

independent of $\mathbf{u}(\cdot)$, $\mathbf{x}_0$, and $\mathbf{z}_0$. Now, let the observer reconstruction error $\boldsymbol{\epsilon}(t)$ be

$$\boldsymbol{\epsilon}(t) = \mathbf{z}(t) - \mathbf{Tx}(t). \tag{4.38}$$

Then the dynamics of the observer error is given by

$$\dot{\boldsymbol{\epsilon}} = \dot{\mathbf{z}} - \mathbf{T}\dot{\mathbf{x}}. \tag{4.39}$$

From (4.8), (4.35)–(4.37), and (4.39) can be written as

$$\dot{\boldsymbol{\epsilon}} = (\mathbf{D}^0 + \sum_{k=1}^{m} \mathbf{D}^k u_k)\boldsymbol{\epsilon} - (\mathbf{TA} - \mathbf{D}^0\mathbf{T} - \mathbf{E}^0\mathbf{C})\mathbf{x} - \sum_{k=1}^{m} [\mathbf{TN}_k - \mathbf{D}^k\mathbf{T} - \mathbf{E}^k\mathbf{C})u_k\mathbf{x} + (\mathbf{G} - \mathbf{TB})\mathbf{u}. \tag{4.40}$$

Equation (4.40) can be reduced to the homogeneous equation

$$\dot{\boldsymbol{\epsilon}} = \mathbf{D}^0\boldsymbol{\epsilon}(t) \tag{4.41}$$

provided that the observer matrices **T**, **D**, **E**, and **G** satisfy

$$\mathbf{D}^k = \mathbf{0}, \qquad k = 1, \ldots, m, \tag{4.42}$$

$$\mathbf{TN}_k = \mathbf{E}^k\mathbf{C}, \qquad k = 1, \ldots, m, \tag{4.43}$$

$$\mathbf{TA} - \mathbf{D}^0\mathbf{T} = \mathbf{E}^0\mathbf{C}, \tag{4.44}$$

$$\mathbf{G} = \mathbf{TB}. \tag{4.45}$$

Also from (4.24),

$$\mathbf{K}\begin{bmatrix}\mathbf{C}\\ \mathbf{T}\end{bmatrix}^{-1} = [\mathbf{V} \quad \mathbf{P}]. \tag{4.46}$$

Furthermore, if $\mathbf{D}^0$ is Hurwitz, $\boldsymbol{\epsilon}(t)$ goes to zero as t goes to infinity.

Then it can be seen that (4.36)–(4.37) is a q-dimensional observer for $\mathbf{Kx}$ of bilinear system (4.8) and (4.35) if and only if there exists a $q \times n$ matrix **T** satisfying (4.42)–(4.46) with $\mathbf{D}^0$ Hurwitz.

It is seen that (4.44)–(4.46) are the same as these conditions, and the two extra conditions exist. Without much loss of generality, bilinear system (4.8) and (4.35) can be written as

$$\dot{\mathbf{x}} = \begin{bmatrix}\mathbf{A}_{11} & \mathbf{A}_{12}\\ \mathbf{A}_{21} & \mathbf{A}_{22}\end{bmatrix}\mathbf{x} + \sum_{k=1}^{m}\begin{bmatrix}\mathbf{N}_{11}^k & \mathbf{N}_{12}^k\\ \mathbf{N}_{21}^k & \mathbf{N}_{22}^k\end{bmatrix}u_k\mathbf{x} + \begin{bmatrix}\mathbf{B}_1\\ \mathbf{B}_2\end{bmatrix}\mathbf{u}, \tag{4.47}$$

$$\mathbf{y} = [\mathbf{I}_r \quad \mathbf{0}]\mathbf{x}, \tag{4.48}$$

where $\mathbf{N}_{11}^k \in \mathbb{R}^{p\times p}$, $\mathbf{N}_{12}^k \in \mathbb{R}^{p\times(n-p)}$, $\mathbf{N}_{22}^k \in \mathbb{R}^{(n-p)\times(n-p)}$, $\mathbf{B}_1 \in \mathbb{R}^{p\times m}$, $\mathbf{B}_2 \in \mathbb{R}^{(n-p)\times m}$, and I_r is the r-dimensional identity matrix.

Also, we can decompose **T** as

$$\mathbf{T} = [\mathbf{T}_1 \quad \mathbf{T}_2]. \tag{4.49}$$

where $\mathbf{T}_1 \in \mathbb{R}^{(n-p)\times p}$ and $\mathbf{T}_2 \in \mathbb{R}^{(n-p)\times(n-p)}$.

Let $x_1 = i_a$, $x_2 = \theta$, $x_3 = \omega$, $u = i_e$, $v = v_a$. Then

$$\begin{bmatrix} \dot{\mathbf{x}}_1 \\ \dot{\mathbf{x}}_2 \\ \dot{\mathbf{x}}_3 \end{bmatrix} = \begin{bmatrix} \dfrac{-R_a}{L_a} & 0 & 0 \\ 0 & 0 & 1 \\ 0 & 0 & \dfrac{-D}{J} \end{bmatrix} \begin{bmatrix} x_1 \\ x_2 \\ x_3 \end{bmatrix} + \begin{bmatrix} 0 & 0 & \dfrac{-K_v}{L_a} \\ 0 & 0 & 0 \\ \dfrac{K_v}{J} & 0 & 0 \end{bmatrix} \begin{bmatrix} x_1 \\ x_2 \\ x_3 \end{bmatrix} u + \begin{bmatrix} \dfrac{1}{L_a} \\ 0 \\ 0 \end{bmatrix} v, \tag{4.58}$$

$$\mathbf{y} = \begin{bmatrix} 1 & 0 & 0 \\ 0 & 1 & 0 \end{bmatrix} \begin{bmatrix} x_1 \\ x_2 \\ x_3 \end{bmatrix}. \tag{4.59}$$

This system is a standard form (4.56) with $n = 3$, $m = 2$, and $p = 2$. We have adapted the motor to a bilinear control system.

To find $\mathbf{L}$, let $\mathbf{L}$ be $\mathbf{p}_1\mathbf{v}_1^{\mathrm{T}}$, a basis of null $[\tilde{\mathbf{N}}_{12}^1]\mathbf{v}_1$, obtained as

$$\begin{bmatrix} \dfrac{-K_v}{L_a} & 0 \end{bmatrix} \mathbf{v}_1 = 0 \Rightarrow \mathbf{v}_1 = [0 \quad 1]^{\mathrm{T}}.$$

Hence

$$D^0 = \tilde{\mathbf{A}}_{22} - p_1\mathbf{v}_1^{\mathrm{T}}\tilde{A}_{12} = \frac{-D}{J} - p_1.$$

Since D^0 must have negative-real eigenvalues, we can choose p_1 such that $-D/J - p_1 < 0$. Therefore, the resulting reduced-order bilinear observer is given by

$$\dot{z} = -\left(\frac{D}{J} + p_1\right)z + \begin{bmatrix} 0 & -p_1\left(\dfrac{D}{J} + p_1\right) \end{bmatrix} \mathbf{y} + \begin{bmatrix} \dfrac{K_v}{J} & 0 \end{bmatrix} \mathbf{y} i_e, \tag{4.60}$$

$$\hat{\mathbf{x}} = \begin{bmatrix} 0 \\ 0 \\ 1 \end{bmatrix} z + \begin{bmatrix} 1 & 0 \\ 0 & 1 \\ 0 & p_1 \end{bmatrix} \mathbf{y}. \tag{4.61}$$

The observer (4.61) was simulated using an

Input u:	$A + B \sin(2\pi Ct)$ (where A, B, and C are constants which may take on a range of values)
Input v:	10 V fixed

The simulation of Fig. 4.4 compares x and $\hat{x}$ and shows that the observer has good performance with small dynamic error for different A, B, and C.

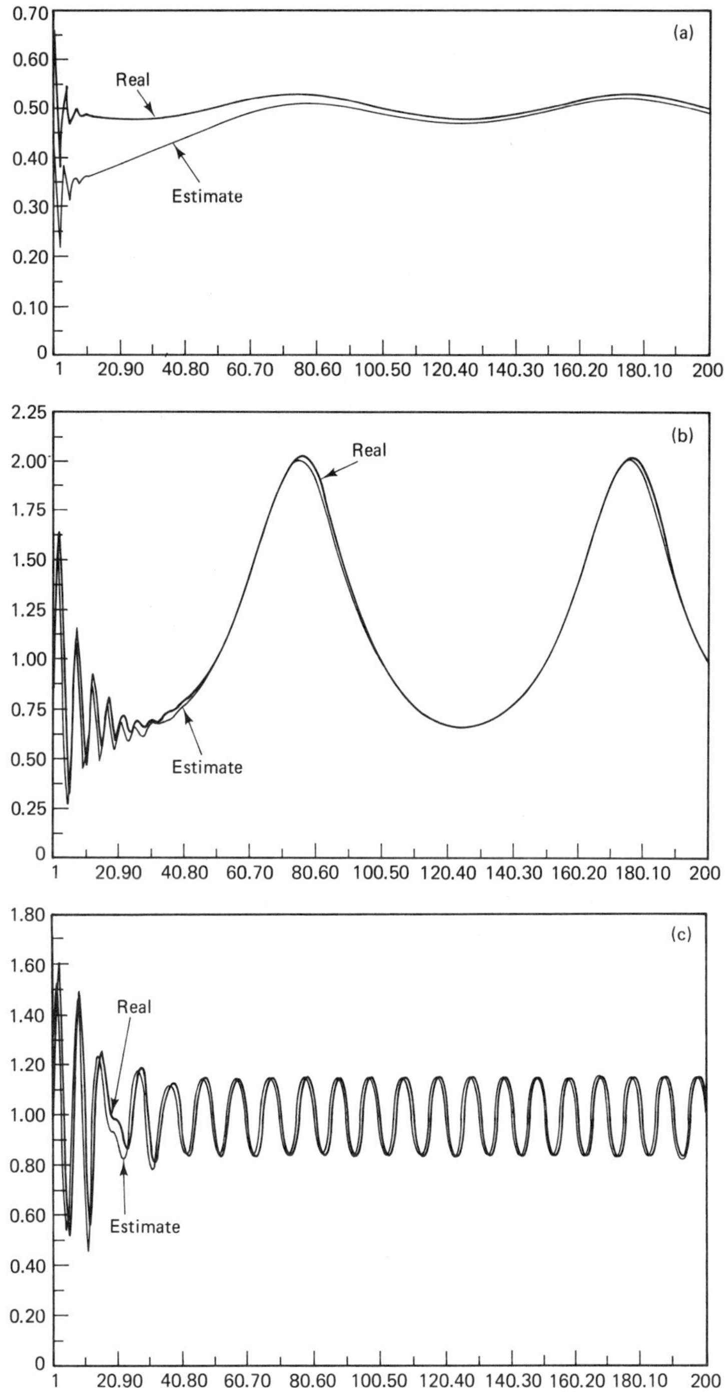

Figure 4.4 Motor observer performance for Example 4.8: (a) $u_1 = 2 + 0.1 \sin(2\pi t)$; (b) $u_1 = 1 + 0.5 \sin(2\pi t)$; (c) $u_1 = 1 + 0.1 \sin(20\pi t)$.

4.4 COMBINATORIAL PROPERTIES

Again consider the scalar-control bilinear system (3.21), which is repeated for convenience as

$$\frac{d\mathbf{x}}{dt} = \mathbf{Ax} + [\mathbf{Nx} + \mathbf{b}]u, \tag{4.62}$$

where $\mathbf{x} \in \mathbb{R}^n$, $u \in \mathbb{R}$; $\mathbf{A}$, $\mathbf{N}$, and $\mathbf{b}$ are constant and of appropriate dimensions. Here the scalar case is assumed to simplify the algebra, but the results do generalize to the vector, $\mathbf{u} \in \mathbb{R}^m$, case.

If (4.62) has an added constant vector $\mathbf{d}$, the system is sometimes called an affine bilinear system, but this definition varies in the literature. Now we have

$$\frac{d\mathbf{x}}{dt} = \mathbf{Ax} + [\mathbf{Nx} + \mathbf{b}]u + \mathbf{d}, \tag{4.63}$$

where $\mathbf{d} \in \mathbb{R}^n$. Sometimes it is convenient to transform (4.63), or (4.62), where $\mathbf{d} = \mathbf{0}$, into a homogeneous BLS

$$\frac{d\bar{\mathbf{x}}}{dt} = \bar{\mathbf{A}}\bar{\mathbf{x}} + \bar{\mathbf{N}}\bar{\mathbf{x}}u, \tag{4.64}$$

where $\bar{\mathbf{x}}^{\mathrm{T}} = [1 \quad \mathbf{x}^{\mathrm{T}}]$,

$$\bar{\mathbf{A}} = \begin{bmatrix} \mathbf{0} & \mathbf{0} \\ \mathbf{d} & \mathbf{A} \end{bmatrix}, \qquad \bar{\mathbf{N}} = \begin{bmatrix} \mathbf{0} & \mathbf{0} \\ \mathbf{b} & \mathbf{N} \end{bmatrix}.$$

Consider the connections of two affine bilinear systems

$$\begin{aligned} \frac{d\mathbf{x}_1}{dt} &= \mathbf{A}_1\mathbf{x}_1 + \mathbf{N}_1\mathbf{x}_1u + \mathbf{b}_1u + \mathbf{d}_1, \\ \frac{d\mathbf{x}_2}{dt} &= \mathbf{A}_2\mathbf{x}_2 + \mathbf{N}_2\mathbf{x}_2u + \mathbf{b}_2u + \mathbf{d}_2. \end{aligned} \tag{4.65}$$

1. The parallel connection $\mathbf{y} = \mathbf{x}_1 + \mathbf{x}_2$, as shown in Fig. 4.5, yields

$$\frac{d\mathbf{y}}{dt} = \frac{d\mathbf{x}_1}{dt} + \frac{d\mathbf{x}_2}{dt} = (\mathbf{A}_1 - \mathbf{A}_2)\mathbf{x}_1 + \mathbf{A}_2\mathbf{y} + (\mathbf{N}_1 - \mathbf{N}_2)\mathbf{x}_1u + \mathbf{N}_2\mathbf{y}u$$
$$+ (\mathbf{b}_1 + \mathbf{b}_2)u + (\mathbf{d}_1 + \mathbf{d}_2).$$

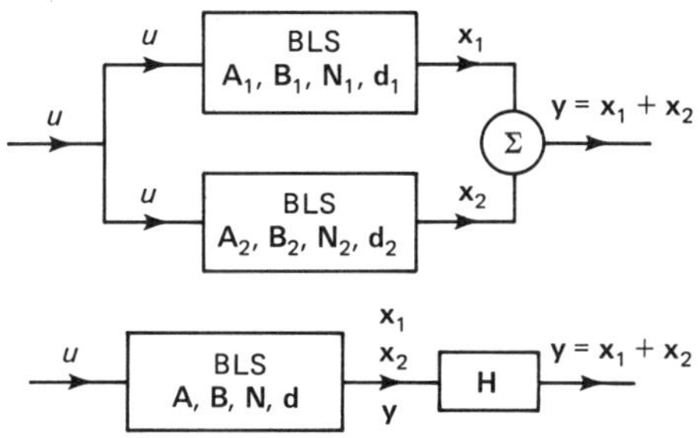

Figure 4.5 Parallel-addition equivalence of BLS.

Then

$$\begin{bmatrix} \dfrac{d\mathbf{x}_1}{dt} \\ \dfrac{d\mathbf{y}}{dt} \end{bmatrix} = \begin{bmatrix} \mathbf{A}_1 & 0 \\ \mathbf{A}_1 - \mathbf{A}_2 & \mathbf{A}_2 \end{bmatrix}\begin{bmatrix} \mathbf{x}_1 \\ \mathbf{y} \end{bmatrix} + \begin{bmatrix} \mathbf{N}_1 & 0 \\ \mathbf{N}_1 - \mathbf{N}_2 & \mathbf{N}_2 \end{bmatrix}\begin{bmatrix} \mathbf{x}_1 \\ \mathbf{y} \end{bmatrix} u + \begin{bmatrix} \mathbf{b}_1 \\ \mathbf{b}_1 + \mathbf{b}_2 \end{bmatrix} u + \begin{bmatrix} \mathbf{d}_1 \\ \mathbf{d}_1 + \mathbf{d}_2 \end{bmatrix} \tag{4.66}$$

or

$$\begin{bmatrix} \dfrac{d\mathbf{x}_1}{dt} \\ \dfrac{d\mathbf{x}_2}{dt} \\ \dfrac{d\mathbf{y}}{dt} \end{bmatrix} = \begin{bmatrix} \mathbf{A}_1 & 0 & 0 \\ 0 & \mathbf{A}_2 & 0 \\ \mathbf{A}_1 - \mathbf{A}_2 & 0 & \mathbf{A}_2 \end{bmatrix}\begin{bmatrix} \mathbf{x}_1 \\ \mathbf{x}_2 \\ \mathbf{y} \end{bmatrix} + \begin{bmatrix} \mathbf{N}_1 & 0 & 0 \\ 0 & \mathbf{N}_2 & 0 \\ \mathbf{N}_1 - \mathbf{N}_2 & 0 & \mathbf{N}_2 \end{bmatrix}\begin{bmatrix} \mathbf{x}_1 \\ \mathbf{x}_2 \\ \mathbf{y} \end{bmatrix} u + \begin{bmatrix} \mathbf{b}_1 \\ \mathbf{b}_2 \\ \mathbf{b}_1 + \mathbf{b}_2 \end{bmatrix} u + \begin{bmatrix} \mathbf{d}_1 \\ \mathbf{d}_2 \\ \mathbf{d}_1 + \mathbf{d}_2 \end{bmatrix}. \tag{4.67}$$

Hence the new state $(\mathbf{x}_1, \mathbf{y})$ or $(\mathbf{x}_1, \mathbf{x}_2, \mathbf{y})$ still satisfies a bilinear system.

2. For the Kronecker product $\mathbf{y} = \mathbf{x}_1 \circ \mathbf{x}_2$, where

$$\mathbf{x}_1 \circ \mathbf{x}_2 = \begin{bmatrix} x_{11}\mathbf{x}_2 \\ x_{12}\mathbf{x}_2 \\ \vdots \\ x_{1n}\mathbf{x}_2 \end{bmatrix}, \quad \frac{d(\mathbf{x}_1 \circ \mathbf{x}_2)}{dt} = \frac{d\mathbf{x}_1}{dt} \circ \mathbf{x}_2 + \mathbf{x}_1 \circ \frac{d\mathbf{x}_2}{dt}$$

$$= (\mathbf{A}_1 \circ \mathbf{I} + \mathbf{I} \circ \mathbf{A}_2)(\mathbf{x}_1 \circ \mathbf{x}_2) + (\mathbf{I} \circ \mathbf{d}_2)\mathbf{x}_1 + (\mathbf{d}_1 \circ \mathbf{I})\mathbf{x}_2 + (\mathbf{N}_1 \circ \mathbf{I} + \mathbf{I} \circ \mathbf{N}_2)(\mathbf{x}_1 \circ \mathbf{x}_2)u + (\mathbf{I} \circ \mathbf{b}_2)\mathbf{x}_1 u + (\mathbf{b}_1 \circ \mathbf{I})\mathbf{x}_2 u.$$

Here $\mathbf{y} \in \mathbb{R}^{n^2}$ if $\mathbf{x}_1, \mathbf{x}_2 \in \mathbb{R}^n$; so that a new state $(\mathbf{x}_1, \mathbf{x}_2, \mathbf{y})$ still satisfies a bilinear system (see Fig. 4.6). If $\mathbf{N}_1 = \mathbf{N}_2 = \mathbf{0}$, the original systems are linear or affine linear, but then the new state $(\mathbf{x}_1, \mathbf{x}_2, \mathbf{y})$ does not satisfy a linear system. In fact, it is a simple task to show that a similar transformation can be made for multilinear combinations of the outputs.

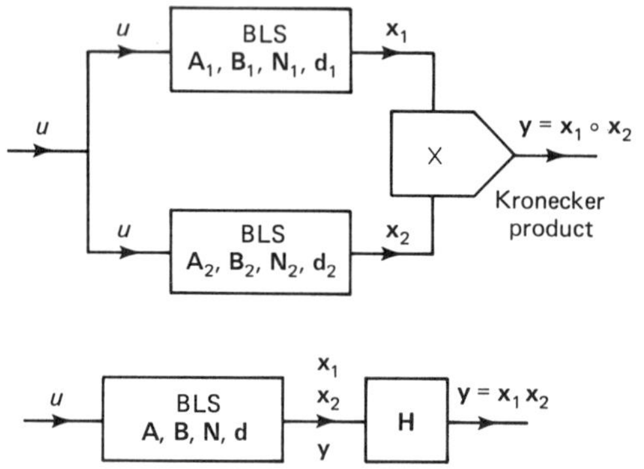

Figure 4.6 Parallel-multiplication equivalence of BLS.

3. Consider the serial connection of Fig. 4.7 such that

$$
\begin{aligned}
\frac{dx_1}{dt} &= a_1x_1 + n_1x_1u + b_1u + d_1, \\
\frac{d\mathbf{x}_2}{dt} &= \mathbf{A}_2\mathbf{x}_2 + \mathbf{N}_2\mathbf{x}_2u + \mathbf{B}_2u + \mathbf{d}_2.
\end{aligned} \tag{4.68}
$$

Only $\mathbf{N}_2 = \mathbf{0}$; $[x_1\ \mathbf{x}_2^T]^T$ satisfies a bilinear system, so if a linear system is followed by a bilinear system, the total system is not bilinear, but if any bilinear system is followed by any linear system, the total system is bilinear. Of course, if linear systems are connected serially, the total system is linear.

For a nonlinear system, suppose that the nonlinearity only comes from the triangular or exponential function of the input, such as $\sin(pu)$, $u \sin(pu)$, e^{pu} . . . ,

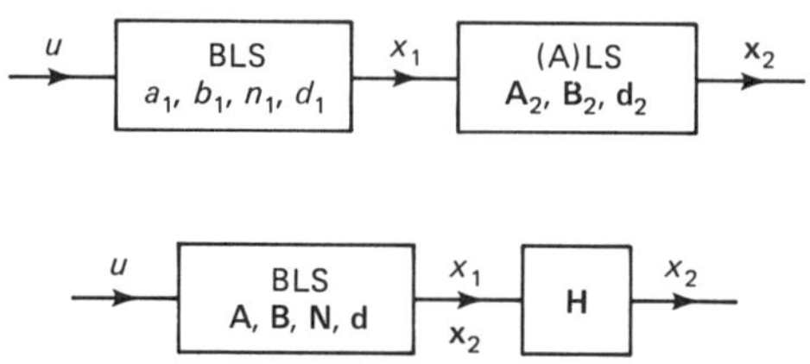

Figure 4.7 Serial linear-system equivalence of BLS.

where p is any constant. Then after appending some new states we can transform the original nonlinear system to a bilinear system with a new input. In fact,

$$\frac{d \sin(pu)}{dt} = p \cos(pu) \frac{du}{dt},$$

$$\frac{d \cos(pu)}{dt} = -p \sin(pu) \frac{du}{dt},$$

$$\frac{de^{pu}}{dt} = pe^{pu} \frac{du}{dt},$$

$$\frac{d\,(u \sin(pu))}{dt} = [\sin(pu) + pu \cos(pu)] \frac{du}{dt},$$

$$\frac{d\,(u \cos(pu))}{dt} = [\cos(pu) - pu \sin(pu)] \frac{du}{dt},$$

$$\frac{d(ue^{pu})}{dt} = (e^{pu} + pue^{pu}) \frac{du}{dt},$$

$$\vdots$$

After appending new states $\sin(pu)$, $\cos(pu)$, e^{pu}, $u \sin(pu)$, . . . , and letting $v = du/dt$ be the new input, this transform results as shown in Fig. 4.8.

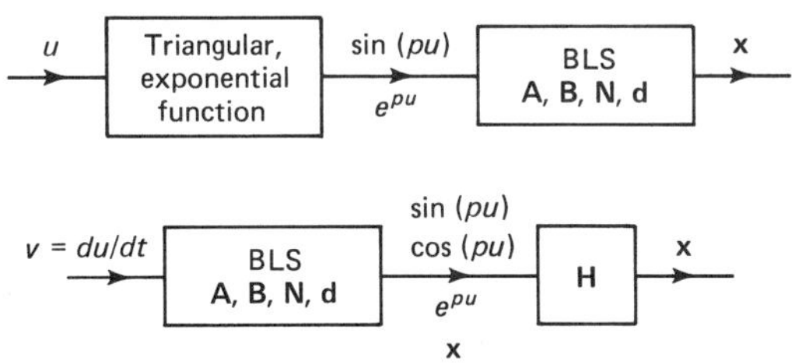

Figure 4.8 Nonlinear input equivalence of BLS.

Of course, superposition does not hold for (4.8) or (4.63), but consider the following properties:

1. If the input $u(t)$ is an affine linear operation such that

$$u(t) = pv(t) + q(t),$$

then the relationship between state $\mathbf{x}(t)$ and $v(t)$ still has the affine BLS structure of (4.63) such that

$$\frac{d\mathbf{x}}{dt} = [\mathbf{A} + \mathbf{N}q(t)]\mathbf{x} + [\mathbf{N}p]\mathbf{x}v(t) + [\mathbf{b}pv(t) + [\mathbf{d} + \mathbf{b}q(t)]. \tag{4.69}$$

2. If the new state vector $\mathbf{y}(t)$ is an (affine) linear static transform of the old state vector $\mathbf{x}(t)$, that is

$$\mathbf{y} = \mathbf{P}\mathbf{x} + \mathbf{q}(t), \tag{4.70}$$

[for linear transform $\mathbf{q}(t) = \mathbf{0}$], where $\mathbf{P}$ is proper and $\dot{\mathbf{q}}$ exists, then the relationship between $\mathbf{y}(t)$ and $u(t)$ still has the BLS structure of (4.63) such that

$$\frac{d\mathbf{y}}{dt} = \mathbf{PAP}^{-1}\mathbf{y} + \mathbf{PNP}^{-1}\mathbf{y}u + [\mathbf{Pb} - \mathbf{PNP}^{-1}\mathbf{q}(t)]u + [\dot{\mathbf{q}}(t) + \mathbf{Pd} - \mathbf{PAP}^{-1}\mathbf{q}(t)]. \tag{4.71}$$

Combining properties 1 and 2, the following useful property results: If the original system is (affine) bilinear, after doing (affine) linear static transforms on state and/or input, respectively, the new state and new input also constitute an (affine) bilinear system as shown in Fig. 4.9. Further structural properties, solutions, and approximate solutions of bilinear systems will be investigated further in Chapter 5.

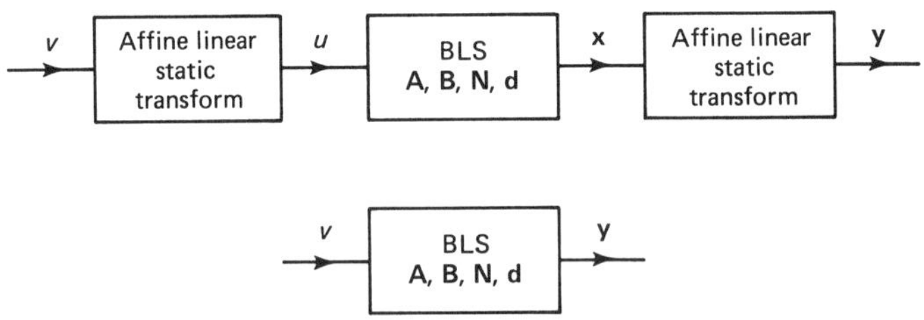

Figure 4.9 Affine property of (affine) BLS.

Example 4.9

Consider the dynamics of a point object moving in a plane with constant speed v and controlled rate of heading $u = \dot{\theta}$ so that in x, y coordinates,

$$\dot{x} = v \cos \theta$$
$$\dot{y} = v \sin \theta.$$

If we let $\mathbf{x}^T = [\sin \theta \quad \cos \theta \quad y \quad x]$, the addended state dynamics becomes the BLS,

$$\dot{\mathbf{x}} = v\mathbf{A}\mathbf{x} + \mathbf{N}\mathbf{x}u,$$

where $\mathbf{x} \in \mathbb{R}^4$, u and v are scalars, and the 4×4 constant matrices are given by

$$\mathbf{A} = \begin{bmatrix} \mathbf{0} & \mathbf{0} \\ \mathbf{I}_2 & \mathbf{0} \end{bmatrix}, \qquad \mathbf{N} = \begin{bmatrix} \mathbf{N}_2 & \mathbf{0} \\ \mathbf{0} & \mathbf{0} \end{bmatrix}.$$

Here $\mathbf{I}_2$ is the 2×2 identity matrix and

$$\mathbf{N}_2 = \begin{bmatrix} 0 & 1 \\ -1 & 0 \end{bmatrix}.$$

4.5 CONCLUDING REMARKS

Bilinear system controllability, observability, and observer design are still fruitful areas of research application. Only a few results are presented here. For a more elegant mathematical treatment the reader is referred to [C6], [B9], and [B10] as well as many of the references provided above. A promising bilinear system observer is derived by means of a Liapunov analysis for convergence in [H7]. It is apparent from the immune model presented in Chapter 2 and the BLS controllability results of this chapter that both parametric control (e.g., by means of cellular stimulation) and additive control (e.g., by a source of stem cells or vaccination of antigen) are important to establish the required controllability of say a rapidly replicating virus. Again, the system judiciously utilizes controlled instability to keep the rapidly expanding alien in check. The stability of a simple immune model, with a bifurcation type of excitation, is given in [C6]. References [C6] and [B9] provide an integration of differential geometry, stability, and controllability of nonlinear systems which is beyond the mission of the present book.

EXERCISES

4.1. Consider the motion of an inertialess object in the xy plane which controls its rate of change in heading $u = \dot{\theta}$.

(a) Describe the object's motion $x(t)$, $y(t)$ in differential equation form.

(b) Define new state variables that make the system bilinear in state and control. Note that for given (or estimated) x, y coordinates, the latter structures may be convenient to ascertain (or estimate) $u(t)$.

(c) Is the bilinear realization one of minimum system order? Explain.

4.2. Consider two BLS

$$\dot{x}_1 = a_1 x_1 + n_1 x_1 u,$$

$$\dot{x}_2 = a_2 x_2 + n_2 x_2 u.$$

Show that by an appropriate selection of state variables, the following operations maintain the BLS structure:

(a) $y = x_1 x_2$

(b) $y = \dfrac{x_1}{x_2}$

(c) $y = x_1^2 x_2$

(d) $y = \cos x_1$

4.3. Compute $\mathbf{x}(t)$ with $\mathbf{x}^T(0) = [1 \quad 1]$ and scalar step input $u(t) = \delta_{-1}(t)$ for the following:

(a) $\dot{x}_1 = (1 + u)(x_1 + x_2)$,
$\dot{x}_2 = (1 + u)x_2$

(b) $\dot{x}_1 = x_1 + x_2 + ux_1$,
$\dot{x}_2 = x_1 + x_2 + ux_2$

4.4. Repeat Exercise 4.3 for general $\mathbf{x}(0)$ and $u(t)$.

4.5. Consider the feedback-controlled BLS with

$$\dot{x}_1 = x_2,$$

$$\dot{x}_2 = x_1 + 2x_2 + ux_1,$$

where

$$u = \begin{cases} -3 & \text{if } s(\mathbf{x}) > 0 \\ 2 & \text{if } s(\mathbf{x}) < 0 \end{cases}$$

and $s(\mathbf{x}) = \alpha x_1 + x_2$.

(a) Sketch the phase-plane portrait if $\alpha = 2$.

(b) Analyze the resulting state of such control if there is imprecise state information to realize $s(\mathbf{x})$.

(c) Repeat parts (a) and (b) if $\alpha = 0.5$.

(d) This problem depicts the potential use of so-called sliding control, which is treated briefly in Chapter 7. Note the different responses available with such control and calculate the precise α that separates the corresponding structures.

(e) Discuss, to the best of your knowledge, the advantages and disadvantages of such sliding control relative to conventional linear feedback (PID) control.

4.6. Analyze controllability of scalar-controlled second-order BLS, $\dot{x}_1 = x_2$, $\dot{x}_2 = a_1 x_1 + a_2 x_2 + nux_2 + bu$ with $|u| \le 1$, where $u(t)$ is a piecewise continuous function. Sketch the possible combinations of linear system state portraits, and show how the eigenvectors limit regions of "controllability."

4.7. Analyze controllability of the following BLS with independent multiplicative control u and additive control v:

$$\dot{x}_1 = u(x_1 - x_2) + v,$$

$$\dot{x}_2 = v.$$

4.8. Explain how a BLS with $|u| \le 1$ can be controllable even if a system eigenvalue passes through the origin as $u(t)$ ranges over the admissible set.

4.9. Derive the following necessary controllability condition for the nth-order BLS with scalar control:

$$\operatorname{rank}[(\mathbf{N}\mathbf{x}_e + \mathbf{b}) \quad (\mathbf{A} + \mathbf{N}u_e)(\mathbf{N}\mathbf{x}_e + \mathbf{b}) \quad \mathbf{N}(\mathbf{N}\mathbf{x}_e + \mathbf{b}) \cdots (\mathbf{A} + \mathbf{N}u_e)^{n-1}(\mathbf{N}\mathbf{x}_e + \mathbf{b}) \cdots \mathbf{N}^{n-1}(\mathbf{N}\mathbf{x}_e + \mathbf{b})] = n$$

at every equilibrium state $\mathbf{x}_e$ with corresponding control u_e.

4.10. Analyze controllability of

$$\dot{x}_1 = x_2 + ux_1 + u,$$
$$\dot{x}_2 = 2ux_2 + u$$

by the result given in Exercise 4.9 and other conditions if required.

4.11. Show that the following conditions are necessary for controllability of the BLS (4.1).

(a) $\mathbf{A}$ has distinct nonzero eigenvalues.

(b) $\mathbf{A}$, $\mathbf{N}_k$ $(k = 1, \ldots, m)$ have no common eigenvectors.

(c) The intersection of the sets

$$V_A \cap V_{N_1} \cdots \cap V_{N_m} = \emptyset,$$

the null set, where

$$V_A = \{\mathbf{x} \,|\, \mathbf{A}\mathbf{X} = 0\},$$
$$V_{N_1} = \{\mathbf{x} \,|\, \mathbf{N}_1\mathbf{x} = -\mathbf{b}_1\},$$
$$\vdots$$
$$V_{N_m} = \{\mathbf{x} \,|\, \mathbf{N}_m\mathbf{x} = -\mathbf{b}_m\},$$

and

$$\mathbf{B} = [b_1 \quad \cdots \quad b_m].$$

4.12. Analyze the controllability of Exercise 4.3.

CHAPTER

5

SYSTEM CHARACTERIZATION and SOLUTIONS

5.1 INTRODUCTION

System structure, which was introduced in Chapter 4, is investigated further in this chapter to characterize nonlinear systems and their solutions. As a consequence of their simple variable structure and their prevalence in applications, bilinear systems are emphasized. First, those cases with exact solutions are studied, which is followed by the Volterra series and Walsh functions. Both of these methods can be used to characterize certain system dynamics and to approximate solutions. Indeed, they do lead to exact solutions if an infinite series is acceptable in general, and even by a finite series in certain specific cases.

5.2 SOLUTION

Existence and uniqueness of solutions is discussed briefly in Vol. 1 Appendix B, but for more details the reader is referred to [C7]. The nonlinear system with a specified $\mathbf{u}(t)$ takes the form

$$\frac{d\mathbf{x}}{dt} = \mathbf{f}(\mathbf{x}, t), \tag{5.1}$$

where $\mathbf{x} \in \mathbb{R}^n$ and $\mathbf{x}(t_0) = \mathbf{x}_0$, and has a unique solution $\mathbf{x}(t)$ if $\mathbf{f}(\mathbf{x}, t)$ satisfies a Lipschitz condition. While this condition is sufficient, it is not necessary. An even stronger sufficient condition, but one that is easier to check, is that the finite derivatives $\partial \mathbf{f}/\partial \mathbf{x}$ and $\partial \mathbf{f}/\partial t$ exist. A weaker condition, continuity of $\mathbf{f}(\mathbf{x}, t)$ in its arguments, ensures the existence of a solution that is sufficient for linear systems

$$\frac{d\mathbf{x}}{dt} = \mathbf{A}(t)\mathbf{x}, \tag{5.2}$$

where $\mathbf{x}(t_0) = \mathbf{x}_0$. But for the linear system, only integrability is required, so that $\mathbf{A}(t)$ piecewise continuous in t is sufficient for a unique solution to exist for (5.2). Similarly, the BLS

$$\frac{d\mathbf{x}}{dt} = \mathbf{A}\mathbf{x} + \sum_{k=1}^{m} \mathbf{N}_k u_k \mathbf{x} + \mathbf{B}\mathbf{u}, \tag{5.3}$$

where $\mathbf{u}(t)$ is a specified piecewise-continuous function, becomes a time-variant system (5.2) with a unique solution.

Again, note the particular BLS cases of interest. First, the scalar control cases ($m = 1$) yields

$$\frac{d\mathbf{x}}{dt} = \mathbf{A}\mathbf{x} + \mathbf{N}u\mathbf{x} + \mathbf{b}u, \tag{5.4}$$

where $\mathbf{A}$, $\mathbf{N}$, and $\mathbf{b}$ are of appropriate dimensions, generally assumed constant but could be time variant for the present discussion. The homogeneous case is given by

$$\frac{d\mathbf{x}}{dt} = (\mathbf{A} + \mathbf{N}u)\mathbf{x}. \tag{5.5}$$

As noted in Section 4.3, (5.4) can be transformed to (5.5) by addending the state vector so that the new $\mathbf{x} \in \mathbb{R}^{n+1}$. However, the loss of controllability and minimum realization of such a representation should be noted.

A slightly more general representation, which was presented in Section 4.3, is the affine bilinear system with

$$\frac{d\mathbf{x}}{dt} = \mathbf{A}\mathbf{x} + \mathbf{N}u\mathbf{x} + \mathbf{b}u + \mathbf{d}, \tag{5.6}$$

where $\mathbf{d} \in \mathbb{R}^n$, and scalar control is assumed for convenience.

For the autonomous case, (5.1), with $\mathbf{f}(\mathbf{x}, t) = \mathbf{f}(\mathbf{x})$, obviously cannot have a solution $\mathbf{x}(t)$ that can reach or leave an equilibrium in finite time if the solution is unique. Similar conclusions can be drawn for various specific cases, but in general the actual solution cannot be found. Accurate approximations may be generated numerically with the aid of a computer. But here the emphasis is on other methods, which include exact solutions in particular cases and analytic approximations such as through the Volterra series or Walsh functions.

5.2.1 Bilinear Systems

For a given input $\mathbf{u}(t)$, the BLS (5.3) is a linear time-variant system with a solution $\mathbf{x}(t)$ that can be expressed in terms of its state transition matrix $\Phi(t, t_0)$, which for the scalar control case (5.4) or (5.5) is the solution to the homogeneous equation

$$\frac{d\Phi}{dt} = [\mathbf{A} + \mathbf{N}u(t)]\Phi. \tag{5.7}$$

Here it is assumed that for a given $u(t)$, $\Phi(t_0, t_0) = \mathbf{I}$ and $\Phi^{-1}(t, t_0)$ exists. Of course, $\mathbf{A}$ and $\mathbf{N}$ can be appropriate time-varying matrix functions at this point.

For a given $u(t)$, it can be seen from the form of solution of (5.4) that

$$\Phi(t, t_0) = \exp \int_{t_0}^{t} [\mathbf{A} + \mathbf{N}u(\tau)]\, d\tau \tag{5.8}$$

if $[\mathbf{A} + \mathbf{N}u(t)]$ and $\exp \int_{t_0}^{t} [\mathbf{A} + \mathbf{N}u(\tau)]\, d\tau$ are commutable. Unfortunately, the solution in general is not that simple.

While the general solution to (5.3), in principle, can be solved in terms of a "homogeneous" solution and a "forced" solution, it is noted that these are not identical to the zero-input response and the zero-state response as for linear systems. In other words, the general solution is not a simple linear superposition of the two solutions. The homogeneous solution actually depends on the input $u(t)$.

This can be seen more clearly by making a transformation from $\mathbf{x}$ to $\mathbf{y}$ such that

$$\mathbf{x}(t) = \Phi(t, t_0)\mathbf{y}(t), \tag{5.9}$$

where $\mathbf{y} \in \mathbb{R}^n$ such that $\mathbf{y}(t_0) = \mathbf{x}_0$. Then differentiation of (5.9) with respect to time to obtain $\dot{\mathbf{x}}(t)$ along with substitution of (5.4) for $\dot{\mathbf{x}}$ yields

$$\dot{\mathbf{y}} = \Phi^{-1}(t, t_0)\mathbf{b}u(t). \tag{5.10}$$

Upon integration of (5.10),

$$\mathbf{y}(t) = \mathbf{x}_0 + \int_{t_0}^{t} \Phi^{-1}(\tau, t_0)\mathbf{B}u(\tau)\, d\tau.$$

Then

$$\mathbf{x}(t) = \Phi(t, t_0)\mathbf{x}_0 + \Phi(t, t_0) \int_{t_0}^{t} \Phi^{-1}(\tau, t_0)\mathbf{B}u(\tau)\, d\tau$$

or

$$\mathbf{x}(t) = \Phi(t, t_0)\mathbf{x}_0 + \Phi(t, t_0) \int_{t_0}^{t} \Phi(t_0, \tau)\mathbf{B}u(\tau)\, d\tau. \tag{5.11}$$

Again, be cautioned, however, that $\Phi(\cdot)$ is actually a function of $u(t)$ as well.

In certain specific cases, such as for (5.8), $\Phi(t, t_0)$ and $\mathbf{x}(t)$ are readily computed. For example, a particular case of (5.8) with $\mathbf{AN} = \mathbf{NA}$. More generally, if the Lie bracket $[\mathbf{A}, \mathbf{N}] = \mathbf{AN} - \mathbf{NA} = c\mathbf{N}$, where c is a constant scalar, it can be shown that with $t_0 = 0$,

$$\mathbf{x}(t) = \exp(\mathbf{A}t) \exp\left[\mathbf{N} \int_{0}^{t} e^{-c\tau}u(\tau)\, d\tau\right]\mathbf{x}_0. \tag{5.12}$$

The solution by means of Lie algebra is derived in Appendix A along with a brief description of Lie algebra.

For the scalar affine bilinear system

$$\frac{dx}{dt} = ax + nux + bu + d, \tag{5.13}$$

the solution may be obtained by direct integration. Consequently,

$$x(t) = \exp\left[at + n \int_{0}^{t} u(\tau)\, d\tau\right] \left\{\int_{0}^{t}\left[d + bu(\tau) \exp\left(-a\tau - n \int_{0}^{\tau} u(\sigma)\, d\sigma\right)\right] d\tau + x_0\right\}. \tag{5.14}$$

If $\mathbf{AB} = \mathbf{BA}$, $\mathbf{BN} = \mathbf{NB}$, and $\mathbf{AN} = \mathbf{NA} + \mathbf{B}$, in (5.3) (i.e., $[\mathbf{A}, \mathbf{N}] = \mathbf{B}$, $[\mathbf{A}, \mathbf{B}] = \mathbf{O}$, $[\mathbf{N}, \mathbf{B}] = \mathbf{O}$), it can be shown (by direct integration or by Lie algebra, Appendix A) that

$$\mathbf{x}(t) = \exp[\mathbf{A}t] \exp\left[\mathbf{N} \int_{0}^{t} u(\tau)\, d\tau\right] \exp\left[\mathbf{B} \int_{0}^{t} \tau u(\tau)\, d\tau\right]\mathbf{x}_0. \tag{5.15}$$

In general, it is possible to obtain the solution in terms of a finite number of multiplications of linear-system solutions if there exists a nonsingular matrix $\mathbf{P}$ such that $\mathbf{PAP}^{-1}$ and $\mathbf{PNP}^{-1}$ are triangular matrices.

5.2.2 Approximate Solutions

For the nonlinear system (5.1) with $t_0 = 0$, $\mathbf{x}(t)$ satisfies

$$\mathbf{x}(t) = \mathbf{x}_0 + \int_0^t \mathbf{f}(\mathbf{x}(\tau), \tau)\, d\tau. \tag{5.16}$$

From (5.16), the classical proof of existence of a unique solution to (5.1), or equivalently (5.16), uses the convergence of Picard's method of successive approximations if a Lipschitz condition is satisfied [C7]. In this manner, construct a sequence of approximations to (5.16) such that

$$\begin{aligned}
\mathbf{x}_0(t) &= \mathbf{x}_0, \\
\mathbf{x}_1(t) &= \mathbf{x}_0 + \int_0^t \mathbf{f}(\mathbf{x}_0(\tau), \tau)\, d\tau, \\
\mathbf{x}_2(t) &= \mathbf{x}_1 + \int_0^t \mathbf{f}(\mathbf{x}_1(\tau), \tau)\, d\tau, \\
&\ \vdots \\
\mathbf{x}_k(t) &= \mathbf{x}_{k-1} + \int_0^t \mathbf{f}(\mathbf{x}_{k-1}(\tau), \tau)\, d\tau, \\
&\ \vdots
\end{aligned} \tag{5.17}$$

Then the Lipschitz condition (Appendix B, Vol. 1) ensures that

$$\lim_{k\to\infty} \|\mathbf{x}_k - \mathbf{x}_{k-1}\| \to 0.$$

For the BLS (5.5), (5.16) becomes

$$\mathbf{x}(t) = \mathbf{x}_0 + \int_0^t [\mathbf{A} + \mathbf{N}u(\tau)]\mathbf{x}(\tau)\, d\tau. \tag{5.18}$$

The approximating sequence is given by

$$\begin{aligned}
\mathbf{x}_0(t) &= \mathbf{x}_0, \\
&\ \vdots \\
\mathbf{x}_k(t) &= \mathbf{x}_{k-1} + \int_0^t [\mathbf{A} + \mathbf{N}u(\tau)]\mathbf{x}_{k-1}(\tau)\, d\tau, \\
&\ \vdots
\end{aligned} \tag{5.19}$$

Here the Lipschitz condition is satisfied if $u(\cdot)$ is bounded on $[0, t]$.

A similar approximating sequence may be generated by assuming that if the input $u(t)$ is multiplied by a constant μ, each successive term in the sequence is multiplied by a μ of one higher order so that $\mathbf{x}(t) \approx \Sigma_i \mu^i \mathbf{x}_i(t)$. Then from (5.18), equating left and right sides of terms for a given power μ yields an infinite cascade of linear systems

$$\begin{aligned}
\frac{d\mathbf{x}_0}{dt} &= \mathbf{A}\mathbf{x}_0, \\
\frac{d\mathbf{x}_1}{dt} &= \mathbf{A}\mathbf{x}_1 + \mathbf{N}\mathbf{x}_0 u, \\
&\vdots \\
\frac{d\mathbf{x}_k}{dt} &= \mathbf{A}\mathbf{x}_k + \mathbf{N}\mathbf{x}_{k-1} u, \\
&\vdots
\end{aligned} \tag{5.20}$$

Here $\mathbf{x}_0(0) = \mathbf{x}_0$ and $\mathbf{x}_i(0) = \mathbf{0}$, $i \neq 0$. Then let $\mu = 1$, so that

$$\mathbf{x}(t) = \sum_{i=0}^{\infty} \mathbf{x}_i(t).$$

The kth approximation of $\mathbf{x}(t)$ is given by

$$\mathbf{x}(t) \approx \sum_{i=0}^{k} \mathbf{x}_i(t).$$

When $u(t)$ is bounded over $[0, t]$, this approximation is convergent. In fact, the solutions of (5.20) are

$$\begin{aligned}
\mathbf{x}_0(t) &= \mathbf{e}^{\mathbf{A}t}\mathbf{x}_0, \\
\mathbf{x}_1(t) &= \mathbf{e}^{\mathbf{A}t} \int_0^t \mathbf{e}^{-\mathbf{A}\tau_1} \mathbf{N} \mathbf{e}^{\mathbf{A}\tau_1} u(\tau_1)\, d\tau_1 \mathbf{x}_0, \\
\mathbf{x}_2(t) &= \mathbf{e}^{\mathbf{A}t} \int_0^t \mathbf{e}^{-\mathbf{A}\tau_2} \mathbf{N} u(\tau_2) \int_0^{\tau_2} \mathbf{e}^{-\mathbf{A}\tau_1} \mathbf{N} \mathbf{e}^{\mathbf{A}\tau_1} u(\tau_1)\, d\tau_1\, d\tau_2\, \mathbf{x}_0, \\
\mathbf{x}_3(t) &= \mathbf{e}^{\mathbf{A}t} \int_0^t \mathbf{e}^{-\mathbf{A}\tau_3} \mathbf{N} u(\tau_3) \int_0^{\tau_3} \mathbf{e}^{-\mathbf{A}\tau_2} \mathbf{N} u(\tau_2) \int_0^{\tau_2} \mathbf{e}^{-\mathbf{A}\tau_1} \mathbf{N} \mathbf{e}^{\mathbf{A}\tau_1} u(\tau_1)\, d\tau_1\, d\tau_2\, d\tau_3\, \mathbf{x}_0, \\
&\vdots
\end{aligned} \tag{5.21}$$

Because $\mathbf{e}^{-\mathbf{A}\tau}\mathbf{N}$ and $\mathbf{e}^{\mathbf{A}\tau}$ are bounded over $[0, t]$, it can be shown that

$$\|\mathbf{x}_1(t)\| < m_1 m_2 t,$$

$$\|\mathbf{x}_2(t)\| < m_1 \frac{1}{2!}(m_2 t)^2,$$

$$\|\mathbf{x}_3(t)\| < m_1 \frac{1}{3!}(m_2 t)^3,$$

$$\vdots$$

where m_1 and m_2 are appropriate constants. The difference between the kth and (k − 1)th approximation is

$$\|x_k(t) - x_{k-1}(t)\| \leq \|x_k(t)\| < m_1 \cdot 1/k!\ (m_2 t)^k$$

and tends to zero as k increases.

For certain special $\mathbf{A}$ and $\mathbf{N}$, only a finite number of nonzero $\mathbf{x}_i(t)$ terms exist, so that after only a finite number the solution is in fact exact. Another method uses Lie algebra. Here the matrices $\mathbf{A}$ and $\mathbf{N}$ and Lie bracket $[\mathbf{A}, \mathbf{N}] = \mathbf{AN} - \mathbf{NA}$, $[\mathbf{A}, [\mathbf{A}, \mathbf{N},]]$, $[\mathbf{A}, [\mathbf{A}, [\mathbf{A}, \mathbf{N}]]]$, . . . consist of a Lie algebra. Let the dimension of the Lie algebra be m, $m \leq n^2$, and matrices $\mathbf{E}_1, \mathbf{E}_2, \ldots, \mathbf{E}_m$ be a set of bases. Assume a solution form

$$\mathbf{x}(t) = \bigcap_{i=1}^{m} \text{Exp}\ (\mathbf{E}_i f_i(t)\mathbf{x}_0),$$

where $f_i(t)$ are uncertain scalar functions with $f_i(0) = 0$. The derivative of $\mathbf{x}(t)$ is

$$\frac{d\mathbf{x}(t)}{dt} = \sum_{j=1}^{m}\left[\bigcap_{i=1}^{j-1} \text{Exp}(\mathbf{E}_i f_i(t))\right]\mathbf{E}_j\left[\bigcap_{i=j}^{m} \text{Exp}\ (\mathbf{E}_i f_i(t))\right]\frac{df_j(t)}{dt}\mathbf{x}_0.$$

Then substitution into (5.5) eventually yields a set of scalar differential equations of uncertain functions $f_i(t)$. Solving $f_i(t)$ means solving $\mathbf{x}(t)$. Unfortunately, $f_i(t)$ are not always solvable; the solvability of $f_i(t)$ depends on the matrices $\mathbf{A}$ and $\mathbf{N}$, and also depends on the care in choosing $\mathbf{E}_i$. An example of this method given in Appendix A yields a solvable case as given above.

Again if there exists a matrix $\mathbf{P}$ such that $\mathbf{PAP}^{-1}$ and $\mathbf{PNP}^{-1}$ are triangular as noted above, the resulting equations in $f_i(t)$ are solvable. Of course, this transformation may be as difficult as the original equations. Although the method provides an exact solution in a finite number of terms for only special cases, such solvable structures may be used to approximate more general unsolvable cases.

5.3 VOLTERRA SERIES

The Volterra series is an intuitively satisfying representation for nonlinear systems which generalizes the concept of the linear system impulse response and its Laplace transform, the transfer function. In this analysis, the linear system solution evolves as a special case of a nonlinear system solution. The concepts of higher-order impulse responses, by means of Volterra kernels and their Laplace transforms, are introduced and are shown to be convenient in the analysis of nonlinear systems under appropriate conditions. Again, however, caution must be observed in attempting to violate the basic assumptions of the methodology. As shown above, for example, the general solution for a nonlinear system, such as the bilinear system, is not a mere linear superposition of zero-state solution and zero-input solution. Only a summary of the Volterra series is presented here. An excellent detailed analysis is provided by [S5] and [R5].

5.3.1 Formulation

Consider a scalar nonlinear system with finite memory such that

1. the input $u(t) = 0$ for $t < 0$, and
2. $u(t - \tau)$, for τ sufficiently large, does not affect the output $y(t)$, and
3. where the required initial conditions on the output are all zero.

In other words, putting it roughly, we consider the zero-state response of physically realizable systems. For example, the pure integrator is not included.

The input–output relationship is derived similar to that for linear systems. First, it is assumed that the input $u(t)$ is approximated by a finite number of finite-dimensional pulses as shown in Fig. 5.1. Each pulse is designated by $u_1, u_2, \ldots, u_N$ starting from $t = t_0$ and numbering back in time to $t = 0$ with pulse width of Δt. As $N \to \infty$ and $\Delta t \to 0$, the approximation of a continuous input becomes exact. The output $y(t)$ may be approximated by a multidimensional Taylor series

$$\tilde{y}(t) = \sum_{i=1}^{N} a_i u_i + \sum\sum_{i,j=1}^{N} a_{ij} u_i u_j + \sum\sum\sum_{i,j,k=1}^{N} a_{ijk} u_i u_j u_k + \ldots, \tag{5.22}$$

where a_i, a_{ij}, a_{ijk} $(i, j, k = 1, \ldots, N)$ are appropriate constants.

The first term in (5.22) represents the linear terms such that with $a_i = h_i \Delta t$,

$$\tilde{y}_1(t_0) = \sum_{i=1}^{N} h_i u_i \, \Delta t.$$

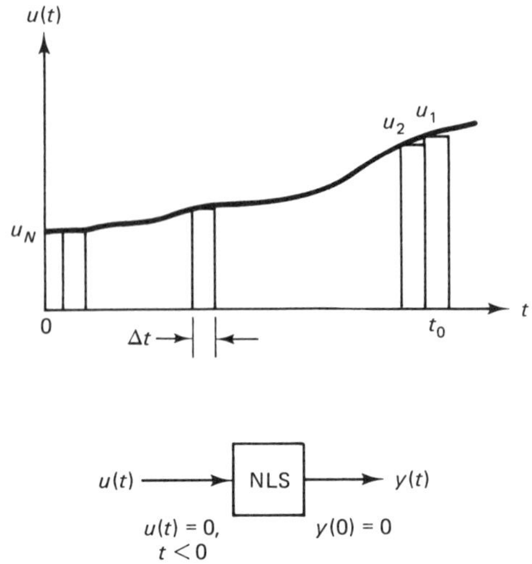

Figure 5.1 Pulse formulation of the input–output relation.

Then in the limit as $N \to \infty$ and $\Delta t \to 0$,

$$y_1(t_0) = \int_0^{t_0} h_i(\tau)u(t_0 - \tau)\, d\tau,$$

or for general time t,

$$y_1(t) = \int_0^{t} h_i(\tau)u(t - \tau)\, d\tau, \tag{5.23}$$

where $h_i(\tau)$ represents the impulse response of the linear portion of the system. That is, the output $y(t)$ is comprised of a weighted superposition of the response to each pulse (weighted delta function)—the convolution integral.

Next, taking the sum of the effects of pulses taken two at a time (with interaction between the pulses for nonlinear systems),

$$\tilde{y}(t_0) = \sum_{i,j=1}^{N}\sum h_{ij}u_iu_j(\Delta t)^2,$$

where $h_{ij} = a_{ij}/(\Delta t)^2$, $i, j = 1, \ldots, N$. Again in the limit as $N \to \infty$ and $\Delta t \to 0$,

$$y_2(t_0) = \int_0^{t_0}\int_0^{t_0} h_2(\tau_1, \tau_2)u(t_0 - \tau_1)u(t_0 - \tau_2)\, d\tau_1\, d\tau_2,$$

or for general time t,

$$y_2(t) = \int_0^t \int_0^t h_2(\tau_1, \tau_2)u(t - \tau_1)u(t - \tau_2)\, d\tau_1 d\tau_2. \tag{5.24}$$

The Kth term is given by

$$y_K(t) = \int_0^t \underset{(K)}{\cdots} \int_0^t h_K(\tau_1, \ldots, \tau_K)u(t - \tau_1) \cdots u(t - \tau_K)\, d\tau_1 \cdots d\tau_K. \tag{5.25}$$

Finally, the output $y(t)$ is given by

$$\begin{aligned} y(t) &= \sum_{i=1}^{\infty} y_i(t) \\ &= \int_0^t h_1(t)u(t - \tau)\, d\tau + \int_0^t \int_0^t h_2(\tau_1, \tau_2)u(t - \tau_1)u(t - \tau_2)\, d\tau_1\, d\tau_2 + \cdots \end{aligned} \tag{5.26}$$

or

$$\begin{aligned} y(t) &= \int_{-\infty}^{\infty} h_1(\tau)u(t - \tau)\delta_{-1}(t - \tau)\, d\tau \\ &+ \int_{-\infty}^{\infty} \int_{-\infty}^{\infty} h_2(\tau_1, \tau_2)u(t - \tau_1)u(t - \tau_2)\delta_{-1}(t - \tau_1)\delta_{-1}(t - \tau_2)\, d\tau_1\, d\tau_2 + \cdots, \end{aligned}$$

where $\delta_{-1}(\cdot)$ is the unit step function. Traditionally, $\delta_{-1}(\cdot)$ is dropped and assumed to be an implicit part of the kernels, so that

$$y(t) = \int_{-\infty}^{\infty} h_1(\tau)u(t - \tau)\, d\tau + \int_{-\infty}^{\infty} \int_{-\infty}^{\infty} h_2(\tau_1, \tau_2)u(t - \tau_1)u(t - \tau_2)\, d\tau_1\, d\tau_2 \cdots \tag{5.27}$$

The following remarks are relevant for this Volterra series representation:

1. The infinite series is an exact description for the input–output description of a continuous nonlinear system.
2. The Volterra series is not an orthogonal series, and consequently there is no assurance that $K + 1$ terms provides a better approximation than does K terms.

3. Despite remark 2, however, the series is convergent under the assumed conditions and a finite number of terms frequently provides a good approximation and sometimes an exact solution.
4. Physical realizability requires that

$$h_K(\tau_1, \ldots, \tau_K) = 0 \qquad \text{for } \tau_i < 0, \quad i = 1, \ldots, K$$

and

$$\lim_{\tau_i \to \infty} h_K(\tau_1, \ldots, \tau_K) = 0.$$

5. Symmetric kernels $h_{sK}(\tau_1, \ldots, \tau_K)$ may be defined so that over all the possible orders of $(\tau_1, \ldots, \tau_K)$,

$$h_{sK}(\tau_1, \ldots, \tau_K) = \frac{1}{K!}[h_K(\tau_1, \ldots, \tau_K) + h_K(\tau_2, \tau_1, \ldots, \tau_K) + \cdots + h_K(\tau_K, \ldots, \tau_1)].$$

5.3.2 Stability

Bounded-input, bounded-output (BIBO) stability has been well established for linear systems. In particular, it is readily shown that for linear time-invariant systems, every bounded input yields a bounded output if and only if its impulse response $h_1(t)$ satisfies

$$\int_{-\infty}^{\infty} |h_1(\tau)|\, d\tau < \infty. \tag{5.28}$$

Therefore, we say that the first-order (linear-system portion) term of the Volterra series is stable if and only if the first-order kernel $h_1(t)$ satisfies the inequality of (5.28).

Similarly, it is seen from (5.24) that with $|u(t)| < M$,

$$|y_2(t)| \leq \int_{-\infty}^{\infty} \int_{-\infty}^{\infty} |h_2(\tau_1, \tau_2)|\, |u(t - \tau_1)|\, |u(t - \tau_2)|\, d\tau_1, d\tau_2$$

or

$$|y_2(t)| < M^2 \int_{-\infty}^{\infty} \int_{-\infty}^{\infty} |h_2(\tau_1, \tau_2)|\, d\tau_1\, d\tau_2. \tag{5.29}$$

Hence, if $h_2(\tau_1, \tau_2)$ satisfies (5.29), the second-order (quadratic) term is BIBO stable. However, it is possible to show by a counterexample that (5.29) is not a necessary condition for such stability.

In general, the scalar nonlinear system is said to be BIBO stable relative to the Kth-order term if

$$\int_{-\infty}^{\infty} \cdots \int_{-\infty}^{\infty} | h_K(\tau_1, \ldots, \tau_K) | \, d\tau_1, \ldots, d\tau_K < \infty. \tag{5.30}$$

The sufficiency of this condition, similar to the second-order case, follows from (5.25) with $|u(t)| < M$, so that

$$| y_K(t) | < M^K \int_{-\infty}^{\infty} \cdots \int_{-\infty}^{\infty} | h_K(\tau_1, \ldots, \tau_K) | \, d\tau_1, \ldots, d\tau_K.$$

But only for the first-order case ($K = 1$) is the condition necessary as well as sufficient.

5.3.3 Computation

The Volterra series expresses the output of a nonlinear system with zero initial conditions by means of an infinite functional series in general. Success of the method depends on the ability to calculate the corresponding Volterra kernels. Two methods are available for such computation: the direct method by substitution of the Volterra series into the system differential or integral equation, and the second method by means of the Laplace transformation and its inverse transformation.

Example 5.1

Consider the linear system with an output quadratic in state as shown in Fig. 5.2. The linearized approximation at some state $\mathbf{x}^*$ results in $\delta y(t) = y(t) - y^* \sim 2x^*\delta x(t)$, where $x^* = \pm(y^*)^{1/2}$ and $\delta x(t) = x(t) - x^*$ with

$$x(t) = \int_{-\infty}^{\infty} h(\tau) u(t - \tau) \, d\tau. \tag{5.31}$$

If $y^* = 0$, the linear approximation $\delta y = 0$.

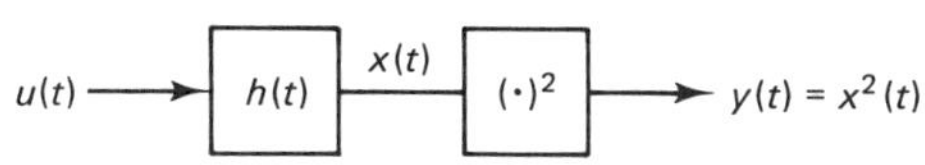

Figure 5.2 Quadratic system of Example 5.1.

However, by direct substitution of (5.31) into $y = x^2$, the exact solution of the output becomes

$$y(t) = \left[\int_{-\infty}^{\infty} h(\tau)u(t - \tau)\, d\tau\right]^2.$$

If the Volterra kernels are desired, continue as follows:

$$\begin{aligned} y(t) &= \int_{-\infty}^{\infty} h(\tau_1)u(t - \tau_1)\, d\tau_1 \int_{-\infty}^{\infty} h(\tau_2)u(t - \tau_2)\, d\tau_2 \\ &= \int_{-\infty}^{\infty}\int_{-\infty}^{\infty} h(\tau_1)h(\tau_2)u(t - \tau_1)u(t - \tau_2)\, d\tau_1\, d\tau_2. \end{aligned}$$

Then comparison with (5.24) yields

$$h_2(\tau_1, \tau_2) = h(\tau_1)h(\tau_2).$$

Only the second-order kernel is nonzero with $y(t) = y_2(t)$. Note that the first-order kernel is zero, which agrees with the linearized perturbation model at $y(0) = 0$.

If $h(t) = Ee^{-\alpha t}\delta_{-1}(t)$, where $\delta_{-1}(t)$ is the unit step function and α is a positive constant, then

$$h_2(\tau_1, \tau_2) = E^2 e^{-\alpha(\tau_1+\tau_2)}\delta_{-1}(\tau_1)\delta_{-1}(\tau_2).$$

Assume that $u(t) = A\delta(t)$ and $x(0) = 0$. Hence, from (5.24) and (5.27),

$$\begin{aligned} y(t) &= A^2 \int_{-\infty}^{\infty}\int_{-\infty}^{\infty} h_2(\tau_1, \tau_2)\delta(t - \tau_1)\delta(t - \tau_2)\, d\tau_1\, d\tau_2, \\ y(t) &= A^2 h_2(t, t) = A^2 E^2 e^{-2\alpha t}\delta_{-1}(t). \end{aligned}$$

Note that $y(t) = y_2(t)$ satisfies (5.28), and therefore the system is BIBO stable.

Now consider the Laplace-transform approach in more detail. The second-order transform of $h_2(\tau_1, \tau_2)$ is given by

$$H_2(s_1, s_2) = \int_{-\infty}^{\infty}\int_{-\infty}^{\infty} h_2(\tau_1, \tau_2)e^{-(s_1\tau_1+s_2\tau_2)}\, d\tau_1 d\tau_2. \tag{5.32}$$

Similarly,

$$y_2(s_1, s_2) = \int_{-\infty}^{\infty}\int_{-\infty}^{\infty} y_2(t_1, t_2)e^{-(s_1 t_1+s_2 t_2)}\, d\tau_1 d\tau_2. \tag{5.33}$$

Then substitution of (5.24) into (5.33) yields

$$y_2(s_1, s_2) = \int_{-\infty}^{\infty}\int_{-\infty}^{\infty}\int_{-\infty}^{\infty}\int_{-\infty}^{\infty} h_2(\tau_1, \tau_2)u(t_1 - \tau_1)u(t_2 - \tau_2)e^{-(s_1t_1+s_2t_2)}\, d\tau_1\, d\tau_2\, dt_1\, dt_2. \tag{5.34}$$

Let $\sigma_1 = t_1 - \tau_1$ and $\sigma_2 = t_2 - \tau_2$ so that (5.34) becomes

$$y_2(s_1, s_2) = \int_{-\infty}^{\infty}\int_{-\infty}^{\infty} h_2(\tau_1, \tau_2)e^{-s_1\tau_1}e^{-s_2\tau_2}\, d\tau_1\, d\tau_2 \int_{-\infty}^{\infty} u(\sigma_1)e^{-s_1\sigma_1}\, d\sigma_1 \int_{-\infty}^{\infty} u(\sigma_2)e^{-s_2\sigma_2}\, d\sigma_2$$

or

$$y_2(s_1, s_2) = H_2(s_1, s_2)u(s_1)u(s_2). \tag{5.35}$$

Similarly, the Kth-order transform of $h_K(\cdot)$ becomes

$$H_K(s_1, \ldots, s_K) = \int_{-\infty}^{\infty}\cdots\int_{-\infty}^{\infty} h_K(\tau_1, \ldots, \tau_K)\exp\left(-\sum_{i=1}^{K} s_i\tau_i\right) d\tau_1, \ldots, d\tau_K \tag{5.36}$$

and

$$y_K(s_1, \ldots, s_K) = H_k(s_1, \ldots, s_K)u(s_1)\cdots u(s_K). \tag{5.37}$$

Now $y_k(t)$ can be obtained from the inverse transform of $y_k(s_1, \ldots, s_K)$ with $t = t_1 = t_2 = \cdots = t_k$. However, sometimes it is less tedious to find the function by an "association of variables," as shown next. For example, consider the second-order case

$$y_2(t_1, t_2) = \frac{1}{(2\pi j)^2}\int_{c_1-j\infty}^{c_1+j\infty}\int_{c_2-j\infty}^{c_2+j\infty} y_2(s_1, s_2)e^{s_1t_1+s_2t_2}\, ds_1\, ds_2. \tag{5.38}$$

Let $t = t_1 = t_2$ and $s = s_1 + s_2$, so that

$$y_2(t) = \frac{1}{(2\pi j)^2}\int_{c_1-j\infty}^{c_1+j\infty}\int_{c_2-j\infty}^{c_2+j\infty} y_2(s - s_2, s_2)\, ds_2 e^{st}\, ds. \tag{5.39}$$

Hence, with

$$y_2(t) = \frac{1}{2\pi j}\int_{c_2-j\infty}^{c_2+j\infty} y_2(s)e^{st}\, ds,$$

it is seen by "association of variables" with (5.39) that

$$y_2(s) = \frac{1}{2\pi j} \int_{c_1 - j\infty}^{c_1 + j\infty} y_2(s - s_2, s_2)\, ds_2. \tag{5.40}$$

Continued application of (5.40) successively to $\{s_2, s_3\}$, then to $\{s_3, s_4\}$, and so on, as was done here for $\{s_1, s_2\}$ reduces $y_k(s_1, \ldots, s_K)$ to $y_k(s)$. Then $y_k(t)$ follows from the conventional one-dimensional Laplace transform.

5.3.4 Systems with Multipliers

Consider Fig. 5.3, where the ouputs of two linear systems are multiplied together to form the input to a third linear system. It is desired to express the output $y(t)$ in terms of the input $u(t)$.

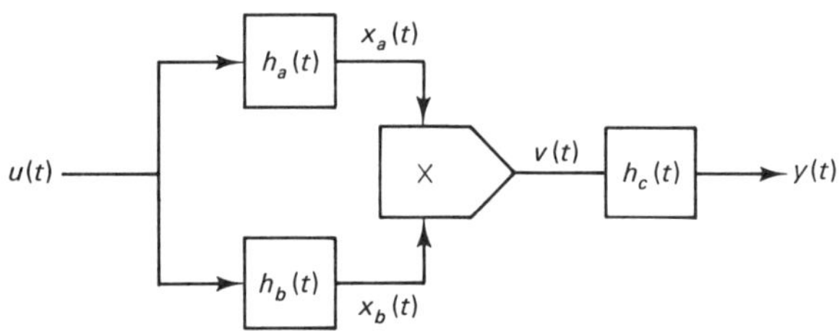

Figure 5.3 Linear system with two outputs multiplied.

The output $y(t)$ may be expressed as a convolution integral of the product of $x_a(t)$ and $x_b(t)$ with the impulse response $h_c(t)$. Thus

$$y(t) = \int_{-\infty}^{\infty} h_c(\sigma) x_a(t - \sigma) x_b(t - \sigma)\, d\sigma, \tag{5.41}$$

where $x_a(\cdot)$ and $x_b(\cdot)$ may be expressed by convolution integrals with their input $u(\cdot)$. Consequently,

$$y(t) = \int_{-\infty}^{\infty} \int_{-\infty}^{\infty} \int_{-\infty}^{\infty} h_c(\sigma) h_a(\sigma_1) h_b(\sigma_2) u(t - \sigma - \sigma_1) u(t - \sigma - \sigma_2)\, d\sigma_1\, d\sigma_2\, d\sigma,$$

or with $\tau_1 = \sigma + \sigma_1$ and $\tau_2 = \sigma + \sigma_2$,

$$y(t) = \int_{-\infty}^{\infty} \int_{-\infty}^{\infty} h_2(\tau_1, \tau_2) u(t - \tau_1) u(t - \tau_2)\, d\tau_1\, d\tau_2, \tag{5.42}$$

where

$$h_2(\tau_1, \tau_2) = \int_{-\infty}^{\infty} h_c(\sigma)h_a(\tau_1 - \sigma)h_b(\tau_2 - \sigma)\, d\sigma. \tag{5.43}$$

Substitution of (5.43) into (5.32) generates

$$H_2(s_1, s_2) = \int_{-\infty}^{\infty}\int_{-\infty}^{\infty}\int_{-\infty}^{\infty} h_c(\sigma)h_a(\tau_1 - \sigma)h_b(\tau_2 - \sigma)e^{-(s_1\tau_1 + s_2\tau_2)}\, d\sigma\, d\tau_1\, d\tau_2.$$

Then let $\beta_1 = \tau_1 - \sigma$ and $\beta_2 = \tau_2 - \sigma$, so that

$$H_2(s_1, s_2) = \int_{-\infty}^{\infty}\int_{-\infty}^{\infty}\int_{-\infty}^{\infty} h_a(\beta_1)h_b(\beta_2)h_c(\sigma)e^{-s_1\beta_1}e^{-s_2\beta_2}e^{-(s_1 + s_2)\sigma}\, d\sigma\, d\beta_1\, d\beta_2$$

or

$$H_2(s_1, s_2) = H_a(s_1)H_b(s_2)H_c(s_1 + s_2). \tag{5.44}$$

Now the Laplace transform of the output may be related to the transform of the input by (5.35), which becomes

$$y(s_1, s_2) = H_2(s_1, s_2)u(s_1)u(s_2)$$

or

$$y(s_1, s_2) = H_a(s_1)H_b(s_1, s_2)u(s_1)u(s_2). \tag{5.45}$$

In summary, the output $y(t)$ may be computed from (5.42), which with causal consideration becomes

$$y(t) = \int_0^t \int_0^t h_2(\tau_1, \tau_2)u(t - \tau_1)u(t - \tau_2)\, d\tau_1\, d\tau_2, \tag{5.46}$$

where $h_2(\tau_1, \tau_2)$ is evaluated by the inverse transform of (5.44). The same result may be obtained by an association of variables as noted in Section 5.3.3.

Now consider Fig. 5.4 with M parallel linear systems whose outputs are multiplied together to form an input $v(t)$ to another linear system with output $y(t)$. Successive application of the previous analysis shows that

$$v(s_1, \ldots, s_M) = \prod_{i=1}^{M} H_i(s_i)u(s_i), \tag{5.47}$$

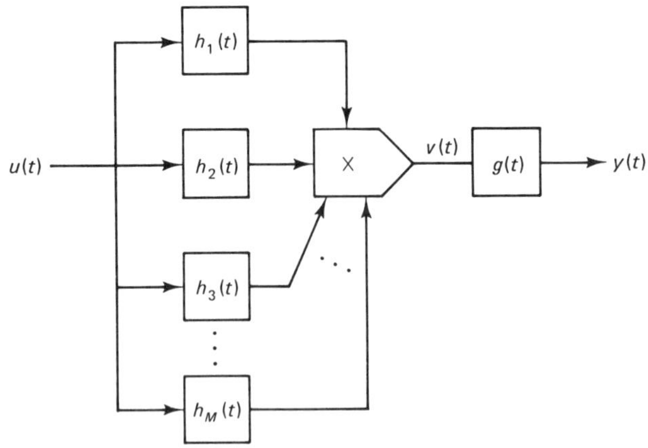

Figure 5.4 Linear system with M outputs multiplied.

and

$$y(s_1, \ldots, s_M) = G(s_1 + s_2 + \cdots + s_M)v(s_1, \ldots, s_M)$$

or

$$y(s_1, \ldots, s_M) = G(s_1 + s_2 + \cdots + s_M) \prod_{i=1}^{M} H_i(s_i)u(s_i). \tag{5.48}$$

Then the Laplace transform of the composite-system kernel is given by

$$H(s_1, \ldots, s_M) = G(s_1 + s_2 + \cdots + s_M) \prod_{i=1}^{M} H_i(s_i). \tag{5.49}$$

If two parallel nonlinear systems have outputs added together as shown in Fig. 5.5(a), it is readily seen that the output for the given kernels, $h_a(\cdot)$ and $h_b(\cdot)$, becomes

$$y(t) = \int_0^t \cdots \int_0^t h_a(\tau_1, \ldots, \tau_M)u(t - \tau_1) \cdots u(t - \tau_M)d\tau_1 \cdots d\tau_M + \int_0^t \cdots \int_0^t h_b(\tau_1, \ldots, \tau_N)u(t - \tau_1) \cdots u(t - \tau_N)\, d\tau_1 \cdots d\tau_N. \tag{5.50}$$

If $M = N$, the transform of the composite system kernel is given by

$$H_a(s_1, \ldots, s_M) + H_b(s_1, \ldots, s_M). \tag{5.51}$$

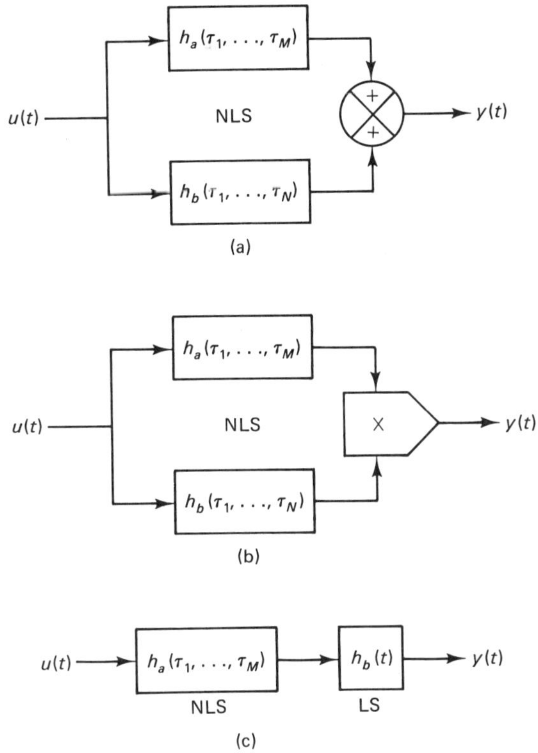

Figure 5.5 Three nonlinear–linear system configurations.

For the product shown in Fig. 5.5(b),

$$y(t) = \int_0^t \cdots \int_0^t h_a(\tau_1, \ldots, \tau_M) h_b(\tau_{M+1}, \ldots, \tau_{M+N}) u(t - \tau_1) \cdots u(t - \tau_{M+N})\, d\tau_1 \cdots d\tau_{M+N}.$$

Here the composite kernel transform is given by

$$H_a(s_1, \ldots, s_M) H_b(s_{M+1}, \ldots, s_{M+N}). \tag{5.52}$$

If the nonlinear system, which is specified by the kernel $h_a(t_1, \ldots, t_M)$, is followed by a linear system with impulse response $h_b(t)$ as shown in Fig. 5.5(c), the output is given by

$$y(t) = \int_0^t \cdots \int_0^t h(\tau_1, \ldots, \tau_M) u(t - \tau_1) \cdots u(t - \tau_M) d\tau_1 \cdots d\tau_M. \tag{5.53}$$

The composite kernel transform becomes

$$H(s_1, \ldots, s_M) = H_a(s_1, \ldots, s_M)H_b(s_1 + \cdots + s_M). \tag{5.54}$$

Example 5.2 [E4]

Consider the nonlinear process of Fig. 5.6. There are two paths from the input $u(t)$ to the output $y(t)$; one is linear with branch transfer function

$$H_1(s) = H_a(s)H_b(s).$$

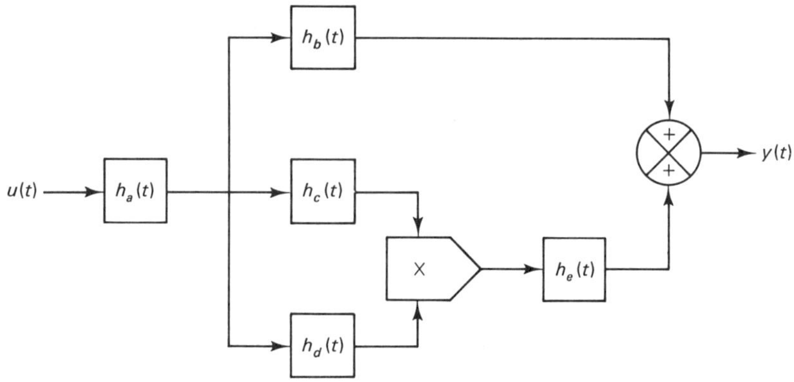

Figure 5.6 Nonlinear system configuration for Example 5.2.

The other is nonlinear with second-order kernel transform given by

$$H_2(s_1, s_2) = H_a(s_1)H_c(s_1)H_a(s_2)H_d(s_2)H_e(s_1 + s_2).$$

The corresponding impulse response $h_1(t)$ and kernel $h_2(t_1, t_2)$ are given by the corresponding inverse transforms. While the frequency-domain representation is quite simple, the time-domain representation is quite complicated with

$$y(t) = \int_0^t h_1(\tau)u(t - \tau)\,d\tau + \int_0^t \int_0^t h_2(\tau_1, \tau_2)u(t - \tau_1)u(t - \tau_2)\,d\tau_1\,d\tau_2.$$

5.3.5 Bilinear Systems

The preceding section shows the generation of the Volterra series for nonlinear systems, which may be decomposed into a finite number of linear systems and multipliers. Here we will see the generalization of this concept to bilinear systems (BLS). As above, first consider the scalar input case for simplicity, so that repeating (5.4) for convenience,

$$\frac{d\mathbf{x}}{dt} = \mathbf{A}\mathbf{x} + \mathbf{N}u\mathbf{x} + \mathbf{b}u. \tag{5.4}$$

To compare like-order terms to generate the Volterra series, multiply the input by a constant μ (as in Section 5.2.2) so that (5.4) becomes

$$\frac{d\mathbf{x}}{dt} = \mathbf{A}\mathbf{x} + \mathbf{N}\mu u\mathbf{x} + \mathbf{b}\mu u \tag{5.55}$$

and

$$\mathbf{x}(t) = \mu\mathbf{x}_1(t) + \mu^2\mathbf{x}_2(t) + \cdots + \mu^K\mathbf{x}_K(t) + \cdots. \tag{5.56}$$

Substitution of (5.56) into (5.55) and a collection of like terms in μ, μ^2, . . . results in the following for $\mu = 1$:

$$\begin{aligned}
\mu^1: \quad \dot{\mathbf{x}}_1 &= \mathbf{A}\mathbf{x}_1 + \mathbf{b}u, \\
\mu^2: \quad \dot{\mathbf{x}}_2 &= \mathbf{A}\mathbf{x}_2 + \mathbf{N}\mathbf{x}_1 u, \\
&\ \vdots \\
\mu^K: \quad \dot{\mathbf{x}}_K &= \mathbf{A}\mathbf{x}_K + \mathbf{N}\mathbf{x}_{K-1}u. \\
&\ \vdots
\end{aligned} \tag{5.57}$$

The corresponding sequence of kernels to (5.57) are given by

$$\begin{aligned}
\mathbf{h}_1(\tau_1) &= \mathbf{e}^{\mathbf{A}\tau_1}\mathbf{b}, \\
\mathbf{h}_2(\tau_1, \tau_2) &= \mathbf{e}^{\mathbf{A}\tau_1}\mathbf{N}\mathbf{e}^{\mathbf{A}(\tau_2-\tau_1)}\mathbf{b}, \\
&\ \vdots \\
\mathbf{h}_K(\tau_1, \ldots, \tau_K) &= \mathbf{e}^{\mathbf{A}\tau_1}\mathbf{N}\mathbf{e}^{\mathbf{A}(\tau_2-\tau_1)} \cdots \mathbf{N}\mathbf{e}^{\mathbf{A}(\tau_K-\tau_{K-1})}\mathbf{b}, \\
&\ \vdots
\end{aligned} \tag{5.58}$$

where $\tau_K \geq \tau_{K-1} \geq \cdots \geq \tau_1 \geq 0$. The $\mathbf{x}_K(t)$ term in general is given by the solution

$$\begin{aligned}
\mathbf{x}_K(\tau_1, \ldots, \tau_K) = \int_0^t \cdots \int_0^t & \mathbf{h}_K(\tau_1, \ldots, \tau_K) \\
& u(t-\tau_1) \cdots u(t-\tau_K)\, d\tau_1 \cdots d\tau_K,
\end{aligned} \tag{5.59}$$

which is the vector generalization of (5.25). Then $\mathbf{x}(t) = \mathbf{x}_1(t) + \mathbf{x}_2(t) + \cdots$.

The reader should note the comparison between (5.58) for the BLS (5.3) with zero initial state and (5.21) for the homogeneous BLS (5.5) with nonzero initial state. The general solution to (5.3) can be computed by the reader as a logical extension.

It is seen from (5.57) and (5.58) that the Volterra series is generated by the previous decomposition into linear systems and multipliers. For example, in Fig. 5.3 with $h_a(t)$ and $h_c(t)$ replaced by $\mathbf{h}_1(t)$ from (5.58) and $\mathbf{e}^{\mathbf{A}t}$, respectively, and $x_b(t)$ replaced by $\mathbf{N}u(t)$, the $y(t)$ becomes $\mathbf{x}_2(t)$ with appropriate algebra. In the frequency domain, $H_a(s)$ is replaced by $\mathbf{H}_1(s) = [\mathbf{I}s - \mathbf{A}]^{-1}\mathbf{b}$, so that

$$\mathbf{H}_2(s_1, s_2) = [\mathbf{I}(s_1 + s_2) - \mathbf{A}]^{-1}\mathbf{N}[\mathbf{I}s_2 - \mathbf{A}]^{-1}\mathbf{b}. \tag{5.60}$$

The third-order kernel is computed by repeating the computation with $\mathbf{h}_2(\tau_1, \tau_2)$ in place of $\mathbf{h}_1(\tau)$, and repeating successively to nest the kernel in the position of the previous kernel synthesizes the higher-order kernels as given by (5.58).

For a vector input, with linear output $\mathbf{y}(t)$, (5.4) becomes

$$\begin{aligned} \frac{d\mathbf{x}}{dt} &= \mathbf{A}\mathbf{x} + \sum_{k=1}^{m} \mathbf{N}_k u_k \mathbf{x} + \mathbf{B}\mathbf{u}, \\ \mathbf{y} &= \mathbf{C}\mathbf{x}, \end{aligned} \tag{5.61}$$

where $\mathbf{x} \in \mathbb{R}^n$, $\mathbf{u} \in \mathbb{R}^m$, $\mathbf{y} \in \mathbb{R}^p$, and $\mathbf{A}$, $\mathbf{B}$, $\mathbf{C}$, $\mathbf{N}_K$ $(k = 1, \ldots, m)$ are constant matrices of appropriate dimensions. Then the state terms of the Volterra series are given as follows:

$$\mathbf{x}_1(t) = \int_0^t \mathbf{e}^{\mathbf{A}\tau_1}\mathbf{B}\mathbf{u}(t - \tau_1)\, d\tau_1$$

or with

$$\begin{aligned} \mathbf{B} &\equiv [\mathbf{b}_1 \quad \mathbf{b}_2 \quad \cdots \quad \mathbf{b}_k \quad \cdots \quad \mathbf{b}_m], \\ \mathbf{x}_1(t) &= \int_0^\infty \mathbf{e}^{\mathbf{A}\tau_1} \sum_{k=1}^{m} \mathbf{b}_k u_k(t - \tau_1)\, d\tau_1. \end{aligned} \tag{5.62}$$

The substitution of (5.62) into the vector equivalence of (5.57) and a suitable change of dummy variables yields

$$\mathbf{x}_2(t) = \int_0^\infty \int_0^\infty \sum_{k_1=1}^{m} \sum_{k_2=1}^{m} \mathbf{e}^{\mathbf{A}\tau_1}\mathbf{N}_{k_1}\mathbf{e}^{\mathbf{A}(\tau_2-\tau_1)}\mathbf{b}_{k_2} u_{k_1}(t - \tau_1) u_{k_2}(t - \tau_2)\, d\tau_1\, d\tau_2, \tag{5.63}$$

and, in general,

$$\mathbf{x}_n(t) = \int_0^\infty \cdots_n \int_0^\infty \sum_{1=k_1,k_2,\ldots,k_n}^{m} \mathbf{e}^{\mathbf{A}\tau_1}\mathbf{N}_{k_1}\mathbf{e}^{\mathbf{A}(\tau_2-\tau_1)}\mathbf{N}_{k_2}\mathbf{e}^{\mathbf{A}(\tau_3-\tau_2)} \cdots \mathbf{N}_{k_{n-1}}\mathbf{e}^{\mathbf{A}(\tau_n-\tau_{n-1})}$$
$$\times\ \mathbf{b}_{k_n}u_{k_1}(t-\tau_1) \cdots u_{k_n}(t-\tau_n)\, d\tau_1 \cdots d\tau_2. \qquad (5.64)$$

Consequently, the state vector $\mathbf{x}$ is

$$\mathbf{x} = \mathbf{x}_1 + \mathbf{x}_2 + \cdots + \mathbf{x}_n + \cdots$$
$$= \sum_{i=1}^{\infty} \int_0^\infty \cdots_i \int_0^\infty \sum_{1=k_1,k_2,\ldots,k_i}^{m} \mathbf{e}^{\mathbf{A}\tau_1}\mathbf{N}_{k_1}\mathbf{e}^{\mathbf{A}(\tau_2-\tau_1)} \cdots \mathbf{N}_{k_{i-1}}\mathbf{e}^{\mathbf{A}(\tau_i-\tau_{i-1})}\mathbf{b}_{k_i}$$
$$\times \left[\prod_{j=1}^{i} u_{k_j}(t-\tau_j)\right] d\tau_1 \cdots d\tau_i. \qquad (5.65)$$

The output vector $\mathbf{y}$ is given by

$$\mathbf{y}(t) = \sum_{i=1}^{m} \int_0^\infty \cdots_i \int_0^\infty \sum_{1=k_1,k_2,\ldots,k_i}^{m} \mathbf{C}\mathbf{e}^{\mathbf{A}\tau_1}\mathbf{N}_{k_1}\mathbf{e}^{\mathbf{A}(\tau_2-\tau_1)} \cdots \mathbf{N}_{k_{i-1}}\mathbf{e}^{\mathbf{A}(\tau_i-\tau_{i-1})}\mathbf{b}_{k_i}$$
$$\times \left[\prod_{j=i}^{i} u_{k_j}(t-\tau_j)\right] d\tau_1 \cdots d\tau_i, \qquad (5.66)$$

which is the Volterra series expansion of the output of the bilinear system of (5.61).

The Volterra kernels are given by

$$h_i^{(k_1,\ldots,k_i)}(\tau_1, \ldots, \tau_i) = \mathbf{C}\mathbf{e}^{\mathbf{A}\tau_1}\mathbf{N}_{k_1}\mathbf{e}^{\mathbf{A}(\tau_2-\tau_1)} \cdots \mathbf{N}_{k_{i-1}}\mathbf{e}^{\mathbf{A}(\tau_i-\tau_{i-1})}\mathbf{b}_{k_i}, \qquad (5.67)$$

$i = 1, 2, \ldots; k_i = 1, 2, \ldots, m;\ \tau_{i+1} \geq \tau_i \geq 0$. They characterize the BLS (5.61) in that knowledge of these kernels provides the means for determining the output corresponding to a given input.

The kernels can be synthesized and are conveniently conceived in Fig. 5.7. Here $\mathbf{H}_1$ is a linear operator characterized by the impulse response $\mathbf{h}(t) = e^{\mathbf{A}t}$. It is readily seen from the above that the Kth-degree kernel is synthesized in a hierarchical manner by successive generations as illustrated by Fig. 5.8 for second- and third-order kernels.

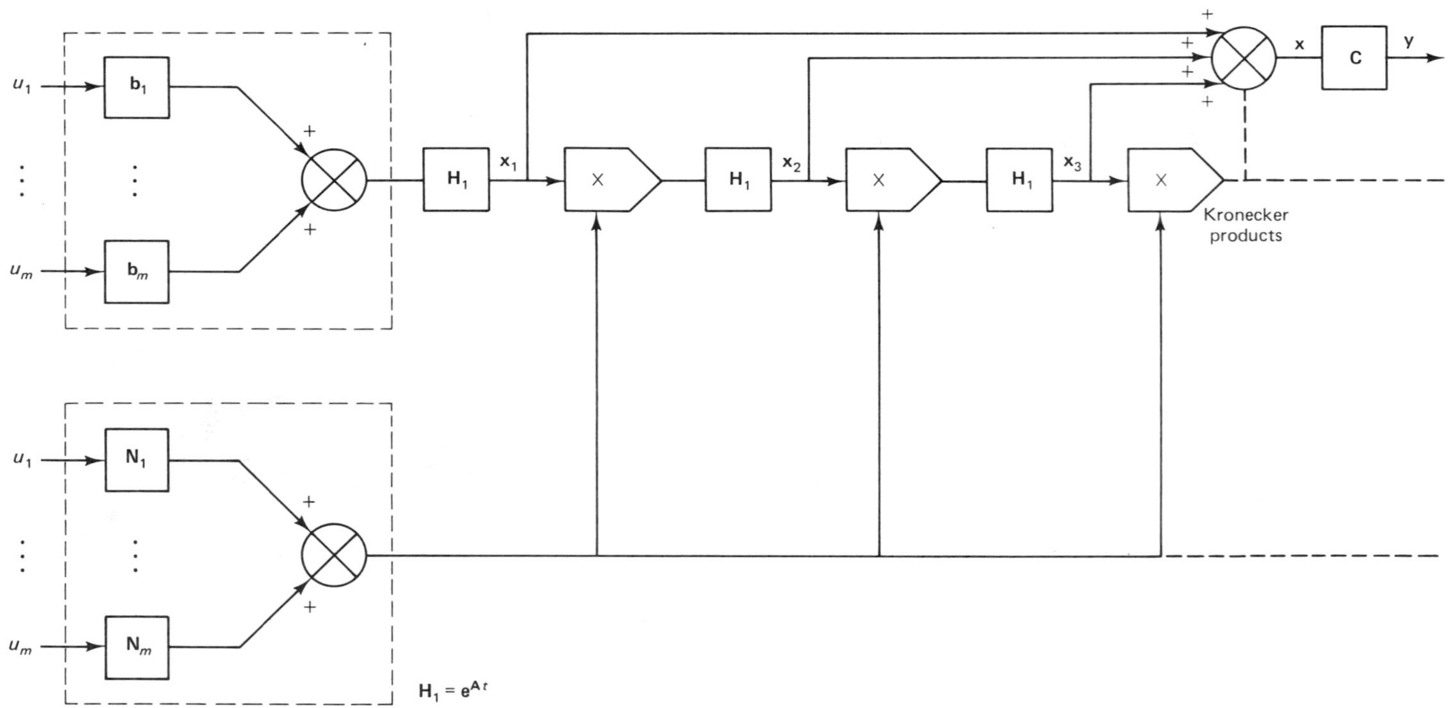

Figure 5.7 BLS Volterra-series diagram; π represents sum of all possible vector–matrix product terms.

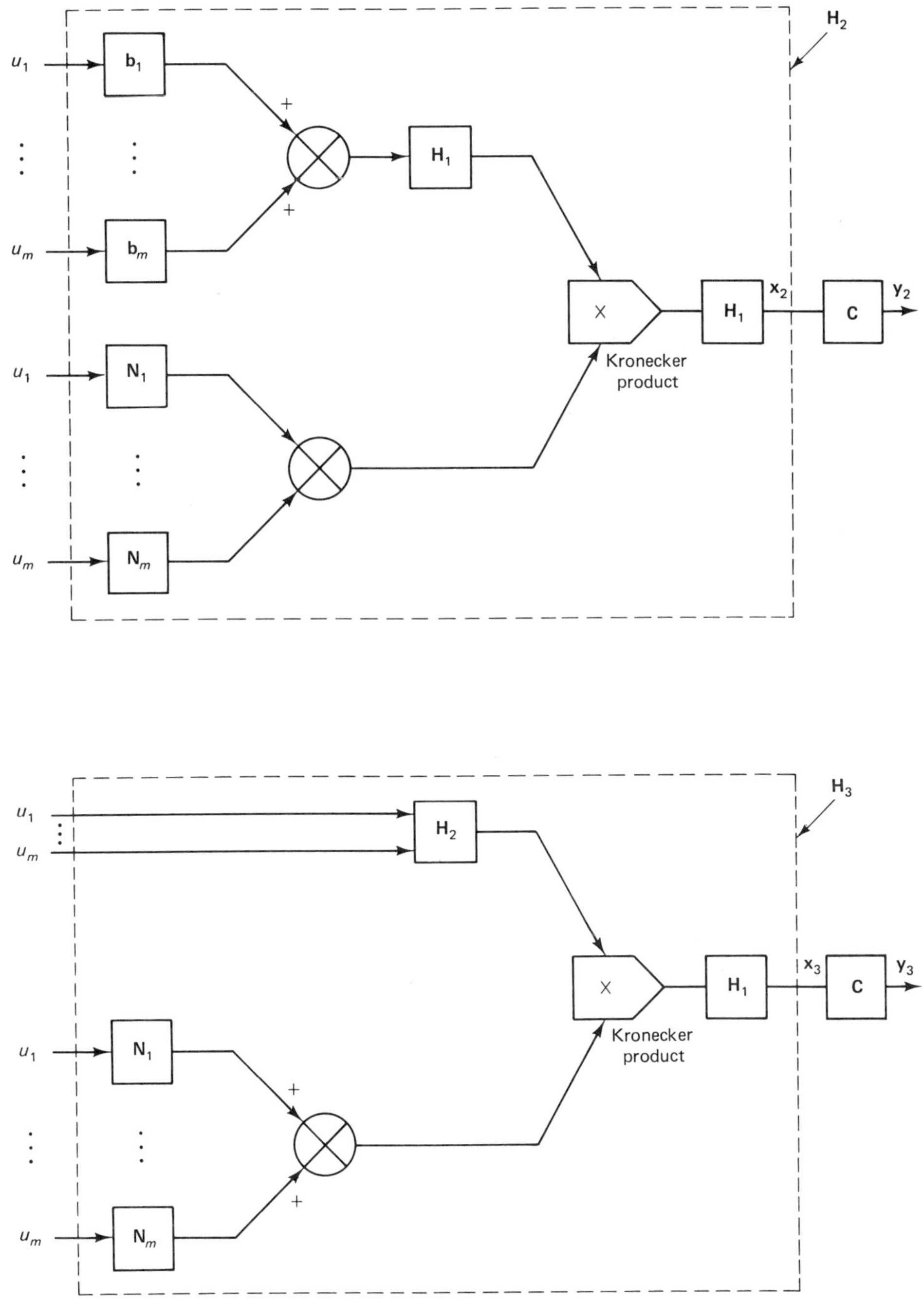

Figure 5.8 Structural synthesis of BLS second- and third-order kernels.

Example 5.3

Consider the BLS

$$\begin{aligned} \dot{x}_1 &= -x_1 + u, \\ \dot{x}_2 &= -2x_2 + ux_1, \\ y &= x_2. \end{aligned} \tag{5.68}$$

(a) Since the first equation of (5.68) is linear, its solution in terms of $u(t)$ may be used to solve the second equation. Apparently, the system is already in the canonical form (triangular matrices **A**, **N**), which yields an exact solution in two steps, as shown by (5.57) and (5.58). By direct integration,

$$x_1(t) = \int_0^\infty e^{-\sigma}u(t-\sigma)\,d\sigma = \int_0^t e^{-(t-\tau)}u(\tau)\,d\tau,$$

$$\begin{aligned} y(t) = x_2(t) &= \int_0^\infty e^{-2\tau}x_1(t-\tau)u(t-\tau)\,d\tau \\ &= \int_0^\infty e^{-2\tau}\left[\int_0^\infty e^{-\sigma}u(t-\tau-\sigma)\,d\sigma\right]u(t-\tau)\,d\tau. \end{aligned}$$

Let $\tau_1 = \tau$ and $\tau_2 = \tau + \sigma$, then

$$y(t) = \int_0^\infty \int_{\tau_1}^\infty e^{-2\tau_1}e^{-(\tau_2-\tau_1)}u(t-\tau_2)u(t-\tau_1)\,d\tau_2\,d\tau_1.$$

Thus

$$h_2(\tau_1, \tau_2) = e^{-(\tau_2-\tau_1)} \tag{5.69}$$

where $\tau_2 \geq \tau_1 \geq 0$ and $h_i(\cdot) = 0$ for $i = 1, 3, 4, \ldots$.

(b) The Laplace transform of the kernel may be computed from its definition (5.32), by (5.60), or from the canonical form of (5.44). Consequently,

$$H_2(s_1, s_2) = \frac{1}{(s_2+1)(s_1+s_2+2)} \tag{5.70}$$

and

$$y(s_1, s_2) = H_2(s_1, s_2)u(s_1)u(s_2).$$

(c) Suppose that $u(t) = \delta_{-1}(t)$, the unit-step function, and $x_1(0) = x_2(0) = y(0) = 0$. Then, from (5.68),

$$x_1(t) = 1 - e^{-t},$$
$$y(t) = x_2(t) = \frac{e^{-2t}}{2} - e^{-t} + \frac{1}{2}. \tag{5.71}$$

Example 5.4

Example 5.3 results in a relatively simple solution with only one kernel required. Again, if the BLS can be transformed to triangular form, the solution can be resolved in terms of a finite number of canonical terms in the Volterra series as generated above. Now consider a more complicated case with

$$\begin{aligned}\dot{x}_1 &= -x_1 + x_2 + u,\\ \dot{x}_2 &= -2x_2 + x_1 + ux_1,\\ y &= x_2.\end{aligned} \tag{5.72}$$

Here

$$\mathbf{A} = \begin{bmatrix} -1 & 1 \\ 1 & -2 \end{bmatrix}$$

with eigenvalues

$$\lambda_{1,2} = \frac{-3 \pm \sqrt{5}}{2}.$$

From the inverse transform of $[\mathbf{I}s - \mathbf{A}]^{-1}$, the state-transition matrix for the linear system portion (first-order kernel) becomes

$$\mathbf{\Phi}(t) = \begin{bmatrix} \frac{5+\sqrt{5}}{10}e^{\lambda_1 t} + \frac{5-\sqrt{5}}{10}e^{\lambda_2 t} & \frac{1}{\sqrt{5}}(e^{\lambda_1 t} - e^{\lambda_2 t}) \\ \frac{1}{\sqrt{5}}(e^{\lambda_1 t} - e^{\lambda_2 t}) & \frac{5-\sqrt{5}}{10}e^{\lambda_1 t} + \frac{5+\sqrt{5}}{10}e^{\lambda_2 t} \end{bmatrix} \tag{5.73}$$

Let $\phi_{ij}(t)$ designate the ijth element of (5.73). Then from (5.67),

$$h_1(\tau_1) = \mathbf{C}^{\mathrm{T}}\mathbf{e}^{\mathbf{A}\tau_1}\mathbf{b} = \phi_{21}(\tau_1), \tag{5.74}$$

for $\tau_1 \geq 0$, or

$$h_1(\tau_1) = \frac{1}{\sqrt{5}}(e^{\lambda_1 \tau_1} - e^{\lambda_2 \tau_1}).$$

Similarly,

$$\begin{aligned} h_2(\tau_1, \tau_2) &= \mathbf{C}^T\mathbf{e}^{\mathbf{A}\tau_1}\mathbf{N}\mathbf{e}^{\mathbf{A}(\tau_2-\tau_1)}\mathbf{b} \\ &= \phi_{22}(\tau_1)\phi_{11}(\tau_2 - \tau_1), \end{aligned} \tag{5.75}$$

$$h_3(\tau_1, \tau_2, \tau_3) = \phi_{22}(\tau_1)\phi_{12}(\tau_2 - \tau_1)\phi_{11}(\tau_3 - \tau_2), \tag{5.76}$$

$$\vdots$$

$$\begin{aligned} h_k(\tau_1, \ldots, \tau_K) &= \phi_{22}(\tau_1)\phi_{12}(\tau_2 - \tau_1)\phi_{12}(\tau_3 - \tau_2) \\ &\quad \cdots \phi_{12}(\tau_{K-1} - \tau_{K-2})\phi_{11}(\tau_K - \tau_{K-1}), \end{aligned} \tag{5.77}$$

where $\tau_K \geq \tau_{K-1} \geq \cdots \geq \tau_1 \geq 0$. $\phi_{ij}(\cdot)$ are appropriate terms in (5.73). The corresponding transforms are given by

$$\mathbf{K}(s) = [\mathbf{I}s - \mathbf{A}]^{-1} = q^{-1}(s)\begin{bmatrix} s+2 & 1 \\ 1 & s+1 \end{bmatrix},$$

$$H_1(s_1) = \mathbf{C}^T\mathbf{K}(s_1)\mathbf{b} = \frac{1}{q(s_1)}, \qquad q(s) = s^2 + 3s + 1, \tag{5.78}$$

$$\begin{aligned} H_2(s_1, s_2) &= \mathbf{C}^T\mathbf{K}(s_1 + s_2)\mathbf{N}\mathbf{K}(s_2)\mathbf{b} \\ &= q^{-1}(s_1 + s_2)[(s_2 + 2)(s_1 + s_2 + 1)]q^{-1}(s_2). \end{aligned} \tag{5.79}$$

$$H_3(s_1, s_2, s_2) = \mathbf{C}^T\mathbf{K}(s_1 + s_2 + s_3)\mathbf{N}\mathbf{K}(s_2 + s_3)\mathbf{N}\mathbf{K}(s_3)\mathbf{b}, \tag{5.80}$$

$$\vdots$$

These two examples show the difference between a BLS with a finite Volterra series (sometimes called a weakly BLS) and a general BLS. While most of the concepts exploited here for BLS do carry over to more general nonlinear-system structures, generally they have not been investigated to the extent that BLS have. Computation of the kernels by Lie algebra, similar to the approach introduced in Section 5.1, is studied in [B11].

Similar to the identification of a linear system impulse response by time-averaged cross-correlation of output with a white-noise input, the higher-order kernels may be evaluated by higher-order cross-correlations. This method with a ternary, pseudorandom input sequence is applied to the nuclear-fission BLS in [B12]. Orthogonal series synthesis by means of Laguerre functions and Wiener functionals are presented along with system identification for white Gaussian random inputs in [S5].

5.4 WALSH FUNCTIONS

Here a finite orthonormal expansion is proposed for the identification or approximate solution of nonlinear systems. Again, bilinear systems receive most of the attention due to their convenient structure, which is conducive to analysis by this class of piecewise-constant functions. The analysis follows that of [K10].

5.4.1 Properties

The bilinear system of (5.61) with scalar $c(t)$ (assumed for convenience) can be represented by

$$\begin{aligned}\mathbf{y}(t) = \mathbf{y}(0) &+ \int_0^t \sum_{i=1}^n \mathbf{a}_i(\tau) y_i(\tau)\, d\tau \\ &+ \int_0^t \sum_{k=1}^m \sum_{i=1}^n \mathbf{n}_{k_i}(\tau) y_i(\tau) u_k(\tau)\, d\tau + \int_0^t \sum_{j=1}^m \mathbf{b}_j(\tau) c(\tau) u_j(\tau)\, d\tau,\end{aligned} \tag{5.81}$$

where $\mathbf{a}_i$, $\mathbf{n}_{k_i}$, and $\mathbf{b}_j$ are the column vectors of matrices $\mathbf{A}$, $\mathbf{N}_k$, and $\mathbf{B}$, respectively, and y_i is the ith component of the vector $\mathbf{y}$. Here it is assumed that the coefficients may be time variant. A method of identifying the $\mathbf{a}_i$, $\mathbf{n}_{k_i}$, and $\mathbf{b}_j$ of (5.81) from input-output measurements is developed employing Walsh functions. Some computational results are presented.

Following [F2] it is convenient to define Walsh functions in terms of Rademacher functions. The Rademacher functions $\{\phi_n\}$ form an incomplete set of orthogonal functions on $[0, 1)$ and are defined by

$$\begin{aligned}\phi_0(t) &= \begin{cases} 1 & 0 \le t < \frac{1}{2} \\ -1 & \frac{1}{2} \le t < 1, \end{cases} \\ \phi_0(t+1) &= \phi_0(t), \\ \phi_n(t) &= \phi_0(2^n t), \qquad n = 1, 2, \ldots,\end{aligned} \tag{5.82}$$

and Walsh functions $\{\psi_n\}$ are then obtained by defining

$$\begin{aligned}\psi_0(t) &= 1, \\ \psi_n(t) &= \phi_{n_1}(t)\phi_{n_2}(t) \cdots \phi_{n_r}(t)\end{aligned} \tag{5.83}$$

for $2n = 2^{n_1} + 2^{n_2} + \cdots + 2^{n_r}$, where the nonnegative integers n_i are uniquely determined by $n_{i+1} < n_i$.

The Walsh functions, being a complete orthonormal set, allow a representation of every piecewise continuous (actually only require Lebesgue-measurable) function $\mathscr{F}(t)$ on [0, 1) in the form

$$\mathscr{F}(t) = a_0\psi_0(t) + a_1\psi_1(t) + a_2\psi_2(t) + \cdots, \tag{5.84}$$

where the coefficients a_n are given by

$$a_n = \int_0^1 \psi_n(t)\mathscr{F}(t)\,dt, \qquad n = 0, 1, 2, \ldots. \tag{5.85}$$

Two properties of Walsh functions particularly attractive for our purposes are the following:

1. The Walsh functions are closed under multiplication, with the multiplication for Walsh functions expressed as

$$\psi_m(t)\psi_n(t) = \psi_{m \oplus n}(t). \tag{5.86}$$

   The sign $\oplus$ represents no-carry modulo-2 addition [H10].
2. The integral of a Walsh function can be represented in terms of Walsh functions. For an integral of the form

$$J_k(t) = \int_0^t \psi_k(\sigma)\,d\sigma, \qquad k = 0, 1, 2, \ldots, \tag{5.87}$$

   with $k = 2^n + k'$, $0 \leq k' \leq 2^n$,

$$J_0(t) = \tfrac{1}{2}\psi_0(t) - \tfrac{1}{4}\sum_{p=0}^{\infty} 2^{-p}\psi_{2^p}(t) \tag{5.88}$$

   and

$$J_k(t) = 2^{-(n+2)}\left\{\psi_k(t) - \sum_{p=1}^{\infty} 2^{-p}\psi_{2^{(n+p)}+k}(t)\right\} \qquad \text{for } k \geq 1.$$

   Equation (5.88) may be manipulated to a more compact form (see [K10]) as

$$\int_0^t \mathbf{\Psi}(\sigma)\,d\sigma = \mathbf{P}\mathbf{\Psi}(t). \tag{5.89}$$

Here $\mathbf{\Psi}(\sigma)$ is a column vector consisting of Walsh functions and the matrix $\mathbf{P}$ is termed "operational matrix." $\mathbf{\Psi}$ and $\mathbf{P}$ in (5.89) are of infinite dimension, but finite-dimensional approximations are obtained by restricting $\boldsymbol{\Psi}$ to contain only 2^n Walsh functions for some integer $n > 1$. The resulting square operational matrix $\mathbf{P}$ is therefore of order $q = 2^n$ and

$$\int_0^t \boldsymbol{\Psi}_{(q)}(\sigma)\, d\sigma \approx \mathbf{P}_{(q)}\boldsymbol{\Psi}_{(q)}(t). \tag{5.90}$$

5.4.2 Parameter Identification

Now a method is developed for identifying the parameter vector-valued functions $\mathbf{a}_i(t)$, $\mathbf{n}_{k_i}(t)$ for $i = 1, 2, \ldots, n$; and $\mathbf{b}_j(t)$ for $j = 1, 2, \ldots, m$ of (5.81).

Employing $q = 2^n$ (for some integer n) Walsh functions in $\boldsymbol{\Psi}$, $\mathbf{a}_i(t)$, $\mathbf{b}_i(t)$, $\mathbf{n}_{k_i}(t)$, and $c(t)$ are approximated by their Walsh function expansions on $[0, 1)$ by

$$\begin{aligned} \mathbf{a}_i(t) &\approx \mathbf{F}_i\boldsymbol{\Psi}_{(q)}(t), & \mathbf{b}_j(t) &\approx \mathbf{H}_j\boldsymbol{\Psi}_{(q)}(t), \\ \mathbf{n}_{k_i}(t) &\approx \mathbf{G}_{k_i}\boldsymbol{\Psi}_{(q)}(t), & c(t) &\approx \mathbf{v}^{\mathrm{T}}\boldsymbol{\Psi}_{(q)}(t), \end{aligned} \tag{5.91}$$

thus reducing the identification problem to identifying $\mathbf{F}_i$, $\mathbf{G}_{k_i}$, and $\mathbf{H}_j$.

The known input $\mathbf{u}$ and the output $\mathbf{y}$ functions similarly are approximated by their Walsh function expansions

$$\begin{aligned} y_i(t) &\approx \mathbf{e}_i^{\mathrm{T}}\boldsymbol{\Psi}_{(q)}(t) = \boldsymbol{\Psi}^{\mathrm{T}}{}_{(q)}(t)\mathbf{e}_i \\ u_k(t) &\approx \mathbf{z}_k^{\mathrm{T}}\boldsymbol{\Psi}_{(q)}(t) = \boldsymbol{\Psi}^{\mathrm{T}}{}_{(q)}(t)\mathbf{z}_k. \end{aligned} \tag{5.92}$$

Substitution of (5.91) and (5.92) into (5.81) leads to

$$\begin{aligned} \mathbf{E}\boldsymbol{\Psi}_{(q)}(t) \approx \mathbf{y}(0) &+ \sum_{i=1}^{n} \mathbf{F}_i \int_0^t \boldsymbol{\Psi}_{(q)}(\tau)\boldsymbol{\Psi}^{\mathrm{T}}{}_{(q)}(\tau)\, d\tau\, \mathbf{e}_i \\ &+ \sum_{k=1}^{m}\sum_{i=1}^{n} \mathbf{G}_{k_i} \int_0^t \boldsymbol{\Psi}_{(q)}(\tau)\boldsymbol{\Psi}^{\mathrm{T}}{}_{(q)}(\tau)\mathbf{e}_i\mathbf{z}_k^{\mathrm{T}}\boldsymbol{\Psi}_{(q)}(\tau)\, d\tau \\ &+ \sum_{k=1}^{m} \mathbf{H}_k \int_0^t \boldsymbol{\Psi}_{(q)}(\tau)\boldsymbol{\Psi}^{\mathrm{T}}{}_{(q)}(\tau)\mathbf{v}\mathbf{z}_k^{\mathrm{T}}\boldsymbol{\Psi}_{(q)}(\tau)\, d\tau, \end{aligned} \tag{5.93}$$

where $\mathbf{E}$ represents the matrix formed by $\mathbf{e}_i^{\mathrm{T}}$ as row vectors.

To render (5.93) useful for computations, it is now shown that the matrix

$$\mathbf{M}_{(q)}(t) = \boldsymbol{\Psi}_{(q)}(t)\boldsymbol{\Psi}^{\mathrm{T}}{}_{(q)}(t) \qquad t \in [0, 1) \tag{5.94}$$

is reducible to a simple algebraic form. $\mathbf{M}_{(q)}$ is a symmetric matrix with $\psi_i(t)\psi_j(t)$ as a typical element $(i, j = 0, 1, \ldots, m - 1)$, and employing the multiplication rule for Walsh functions it can be represented in the form

$$\mathbf{M}_{(q)}(t) = \begin{bmatrix} \psi_0(t) & \psi_1(t) & \cdots & \cdots & \psi_{q-1}(t) \\ \psi_1(t) & \psi_0(t) & \cdots & \cdots & \psi_{q-2}(t) \\ \cdots & & \cdots & \cdots & \cdots \\ \cdots & & \psi_{i\oplus j}(t) & \cdots & \cdots \\ \cdots & & \cdots & \cdots & \cdots \\ \psi_{q-1}(t) & \psi_{q-2}(t) & \cdots & \cdots & \psi_0(t) \end{bmatrix}. \tag{5.95}$$

It is seen that the first column of the matrix $\mathbf{M}_{(q)}(t)$ is $\mathbf{\Psi}_{(q)}(t)$ and the remaining columns are permutations of the first column elements. $\mathbf{M}_{(q)}(t)$ can be represented in the composite matrix notation

$$\mathbf{M}_{(q)}(t) = [\mathbf{\Psi}_{(q)}(t), \mathbf{\Lambda}_1^{(q)}\mathbf{\Psi}_{(q)}(t), \mathbf{\Lambda}_2^{(q)}\mathbf{\Psi}_{(q)}(t), \ldots, \mathbf{\Lambda}_{q-1}^{(q)}\mathbf{\Psi}_{(q)}(t)], \tag{5.96}$$

where, for example, when $q = 4$,

$$\mathbf{\Lambda}_1^{(4)} = \begin{bmatrix} 0 & 1 & 0 & 0 \\ 1 & 0 & 0 & 0 \\ 0 & 0 & 0 & 1 \\ 0 & 0 & 1 & 0 \end{bmatrix}, \quad \mathbf{\Lambda}_2^{(4)} = \begin{bmatrix} 0 & 0 & 1 & 0 \\ 0 & 0 & 0 & 1 \\ 1 & 0 & 0 & 0 \\ 0 & 1 & 0 & 0 \end{bmatrix}, \quad \text{etc.}$$

In general, the $\mathbf{\Lambda}_1^{(q)}$ matrices can be obtained from

$$\mathbf{\Lambda}_i^{(q)} = \begin{bmatrix} \mathbf{\Lambda}_i^{(q/2)} & \mathbf{0}^{(q/2)} \\ \mathbf{0}^{(q/2)} & \mathbf{\Lambda}_i^{(q/2)} \end{bmatrix}$$

and

$$\mathbf{\Lambda}_{i+(q/2)}^{(q)} = \begin{bmatrix} \mathbf{0}^{(q/2)} & \mathbf{\Lambda}_i^{(q/2)} \\ \mathbf{\Lambda}_i^{(q/2)} & \mathbf{0}^{(q/2)} \end{bmatrix}, \qquad q > 1 \tag{5.97}$$

for $i = 0, 1, 2, \ldots, (q/2 - 1)$ and $\mathbf{\Lambda}_0^{(q/2)} = \mathbf{I}_{(q/2)}$.

To exploit fully the simple algebraic structure outlined above in the system representation of (5.93), the following two lemmas are established:

Lemma 1. $\mathbf{\Psi}_{(q)}\mathbf{\Psi}_{(q)}^{\mathrm{T}}\boldsymbol{\alpha} = \mathbf{\Lambda}_\alpha\mathbf{\Psi}_{(q)}$, where $\mathbf{\Lambda}_\alpha = [\boldsymbol{\alpha}, \mathbf{\Lambda}_1^{(q)}\boldsymbol{\alpha}, \ldots, \mathbf{\Lambda}_{q-1}^{(q)}\boldsymbol{\alpha}]$.

Lemma 2. $\mathbf{\Psi}_{(q)}\mathbf{\Psi}_{(q)}^{\mathrm{T}}\boldsymbol{\alpha}\boldsymbol{\beta}^{\mathrm{T}}\mathbf{\Psi}_{(q)} = \mathbf{\Lambda}_\alpha\mathbf{\Lambda}_\beta\mathbf{\Psi}_{(q)}$.

From (5.90) and Lemmas 1 and 2, (5.94) can be written as

$$\mathbf{E}\boldsymbol{\Psi}_{(q)}(t) \approx \mathbf{y}(0) + \sum_{i=1}^{n} \mathbf{F}_i \boldsymbol{\Lambda}_{e_i} \mathbf{P}_{(q)} \boldsymbol{\Psi}_{(q)}(t) + \sum_{k=1}^{m} \sum_{i=1}^{n} \mathbf{G}_{k_i} \boldsymbol{\Lambda}_{e_i} \boldsymbol{\Lambda}_{z_k} \mathbf{P}_{(q)} \boldsymbol{\Psi}_{(q)}(t) + \sum_{k=1}^{m} \mathbf{H}_k \boldsymbol{\Lambda}_v \boldsymbol{\Lambda}_{z_k} \mathbf{P}_{(q)} \boldsymbol{\Psi}_{(q)}(t). \tag{5.98}$$

It is to be noted that the above holds for arbitrary $t \in [0, 1)$ and therefore holds for all t in this interval. It is now a relatively simple computational task to identify $\mathbf{F}_i$, $\mathbf{G}_{k_i}$, and $\mathbf{H}_k$ for known $\mathbf{E}$, $\mathbf{z}_k$, and $\mathbf{v}$ from input–output data. m linearly independent equations that contain all the parameters can be generated from (5.98) by sampling the Walsh functions at $t = (2i - 1)/2q$ for $i = 1, 2, \ldots, q$. These equations can be solved for the unknowns $\mathbf{F}_i$, $\mathbf{G}_{k_i}$, and $\mathbf{H}_K$. Situations may arise where the number of unknowns is greater than q. For example, in the identification, using eight Walsh functions of a second-order time-invariant bilinear system with a two-input vector, the unknown components of matrices $\mathbf{A}$, $\mathbf{N}_k$, and $\mathbf{B}$ total 16. Such situations may be handled by using additional input–output data with a different initial condition or with a different input vector. Another possible solution to such a problem is to increase the number of Walsh functions being used for the identification.

Example 5.5 [K10]

Consider the bilinear system with time-varying parameters

$$\dot{y}(t) = a(t)y(t) + n(t)y(t)u(t) + b(t)u(t). \tag{5.99}$$

With (5.98) corresponding to (5.99),

$$a(t) = \mathbf{f}^{\mathrm{T}}\boldsymbol{\Psi}(t)$$
$$n(t) = \mathbf{g}^{\mathrm{T}}\boldsymbol{\Psi}(t)$$
$$b(t) = \mathbf{h}^{\mathrm{T}}\boldsymbol{\Psi}(t)$$

can be written as

$$\mathbf{e}^{\mathrm{T}}\boldsymbol{\Psi}(t) = y(0) + \mathbf{f}^{\mathrm{T}}\boldsymbol{\Lambda}_e\mathbf{P}\boldsymbol{\Psi}(t) + \mathbf{g}^{\mathrm{T}}\boldsymbol{\Lambda}_e\boldsymbol{\Lambda}_z\mathbf{P}\boldsymbol{\Psi}(t) + \mathbf{h}^{\mathrm{T}}\boldsymbol{\Lambda}_z\mathbf{P}\boldsymbol{\Psi}(t), \tag{5.100}$$

where $\mathbf{e}$ and $\mathbf{z}$ are the Walsh coefficient vectors of $y(t)$ and $u(t)$, respectively. Using q Walsh functions, it is required to find $3q$ components of $\mathbf{f}$, $\mathbf{g}$, and $\mathbf{h}$. Four different inputs $e^{-2t}(0.2)$, $2t(0.1)$, $te^{-t}(4.1)$, and $1/(t + 0.5)(2.6)$ (the numbers in parentheses being the initial conditions), are used in (5.100) to generate $4m$ equations to solve for the $3m$ unknowns. The plots of $a(t)$, $n(t)$, and $b(t)$ for $q = 4$ and $q = 8$ Walsh functions are illustrated in Fig. 5.9.

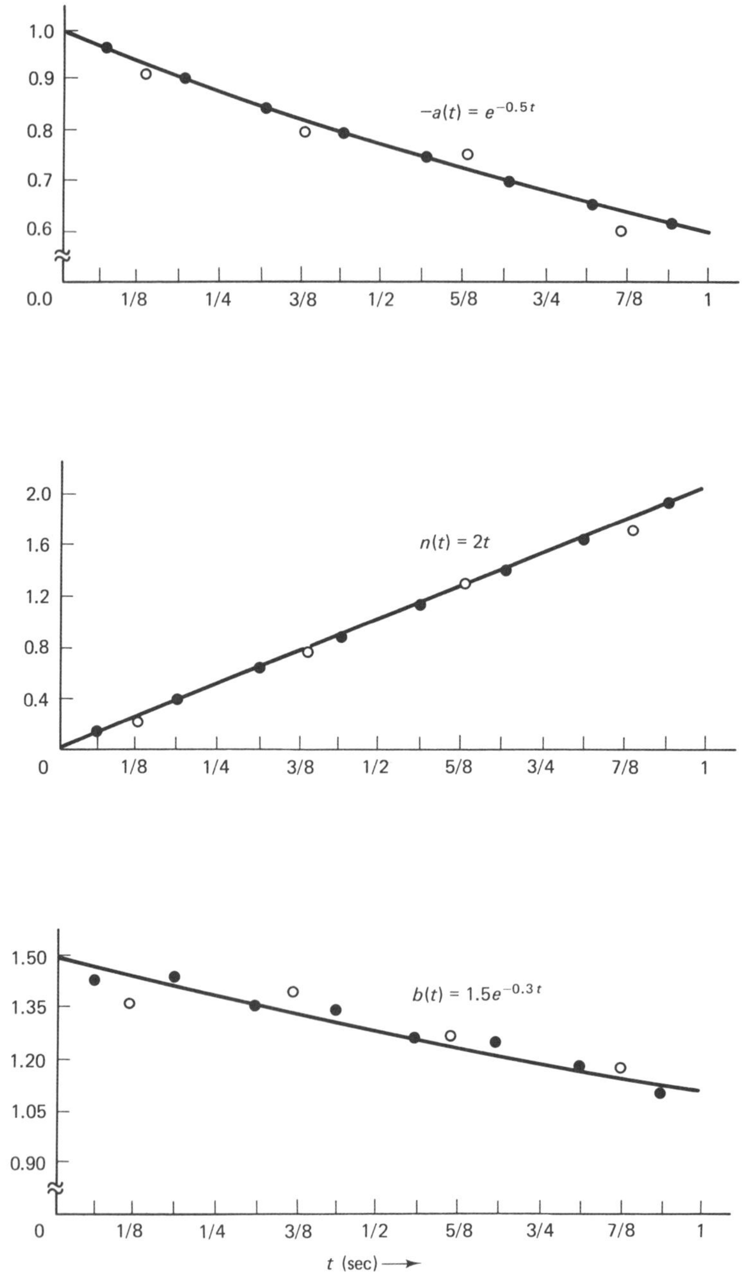

Figure 5.9 Computed results for Example 5.5. Open circles designate four Walsh functions; solid circles designate eight Walsh functions.

Example 5.6 [M14]

Consider the humoral immune system (2.36) with time normalized by definition of

$$\tau = \frac{t - t_0}{t_f - t_0},$$

where t_0 and t_f are the initial and final times of the data set. Then the generation of plasma cells (2.32) is given by

$$\frac{dx_2}{d\tau} = (t_f - t_0)\left[2\alpha u_2 x_1 - \frac{x_2}{\tau_2}\right], \tag{5.101}$$

where $\tau \in (0, 1)$. Then (5.101) may be approximated by

$$\mathbf{e}^{\mathrm{T}}\boldsymbol{\Psi}_{(q)}(\tau) \approx x_2(t_0) - \frac{t_f - t_0}{\tau_2}\mathbf{e}^{\mathrm{T}}\mathbf{P}_{(q)}\boldsymbol{\Psi}_{(q)}(\tau) + 2\alpha(t_f - t_0)\mathbf{f}^{\mathrm{T}}\boldsymbol{\Lambda}_z\mathbf{P}_{(q)}\boldsymbol{\Psi}_{(q)}(\tau),$$

where $\mathbf{z}$, $\mathbf{e}$, and $\mathbf{f}$ are the Walsh coefficient vectors of $x_1(\tau)$, $x_2(\tau)$, and $u_2(\tau)$. Also,

$$\boldsymbol{\Lambda}_z = [\mathbf{z} \quad \boldsymbol{\Lambda}_1^{(q)}\mathbf{z} \quad \cdots \quad \boldsymbol{\Lambda}_{q-1}^{(q)}\mathbf{z}],$$

and $\boldsymbol{\Lambda}_i^{(q)}$ are sparse matrices of ones and zeros given by (5.97).

With $t_0 = 160$ hr, $t_f = 240$ hr, and $q = 8$ Walsh function approximations, four equations are generated for the solution of plasma-cell, generation-rate, and lifetime parameters, α and τ_2. From a computer simulation with $\alpha = 0.0578\ \mathrm{hr}^{-1}$, $\tau_2 = 21$ hr, the estimated values are $\hat{\alpha} = 0.0572\ \mathrm{hr}^{-1}$ and $\hat{\tau}_2 = 20.7$ hr.

To estimate τ_1 in (2.31), with $u_1 = 0$ for $t > 240$ hr and v_1 a constant source rate of stem cells, the procedure yields

$$\mathbf{a}^{\mathrm{T}}\boldsymbol{\Psi}_{(q)}(\tau) \approx x_1(t_0) - \frac{t_f - t_0}{\tau_1}\mathbf{a}^{\mathrm{T}}\mathbf{P}_{(q)}\boldsymbol{\Psi}_{(q)}(\tau) + v_1\mathbf{p}_1^{\mathrm{T}}\boldsymbol{\Psi}_{(q)}(\tau).$$

Here $\mathbf{p}_1^{\mathrm{T}}$ is the first row vector of the matrix $\mathbf{P}_{(q)}$, and $\mathbf{a}$ is the new Walsh coefficient vector of $x_1(\tau)$. With $t_1 = 300$ hr and $t_f = 380$ hr, $\hat{\tau}_1 = 7.15(10)^3$ hr. The actual immunocompetent cell average lifetime for the simulation is $\tau_1 = 7.14(10)^3$ hr.

While the derivation of the proposed method is carried through for bilinear systems, the method can be applied to a very large class of nonlinear systems. For example, the polynomial nonlinear term x^n (x, a scalar) can be handled using Lemmas 1 and 2 repeatedly. Let $x(t) = \mathbf{e}^{\mathrm{T}}\boldsymbol{\Psi}(t)$. Then $x^n(t) = \mathbf{e}^{\mathrm{T}}\boldsymbol{\Lambda}_e^{n-1}\boldsymbol{\Psi}(t)$. Fractional powers in nonlinearities, however, cannot be handled in this framework because Lemmas 1 and 2 do not apply and it is doubtful if analogous results could be obtained to handle such cases. Problems with unknown initial conditions pose no additional difficulties and can be handled in an analogous way.

The accuracy of the Walsh-function method in the computational examples is quite good. Moreover, since Walsh functions are relatively easy to handle in computations, the proposed technique is quite simple to implement and has, therefore, obvious advantages in practical situations. The only computational disadvantage is that 2^n (n an integer) Walsh functions have to be employed, and as a consequence the computer storage requirements for higher-dimensional problems might become excessive if more than 8 to 16 functions are used. Acceptable parameter estimates can be obtained even in cases where the bilinear system is subject to modeling and measurement errors [K10].

5.5 CONCLUDING REMARKS

Within each BLS itself there evolves a certain hierarchical structure that is associated with a "canonical" decomposition such as convenient for its Volterra representation (5.57). Here $\mathbf{x}_i$, $i = 1, \ldots,$ correspond to the terms in the Volterra series, and the corresponding kernels are generated as a "nesting" of linear-system impulse responses according to (5.58). The Volterra series for BLS may sometimes be approximated by a finite number of terms (5.65), and in the case of so-called weakly BLS, may be exactly represented by a finite number or decomposed into a finite number of linear systems with states or outputs multiplied together to form inputs to successive linear systems according to (5.57). Consequently, a physical system approximated (or given) by a finite hierarchical structure associated with (5.57) may sometimes be conveniently analyzed by linear system theory.

If BLS (5.61) can be transformed to a "canonical" form with $\mathbf{A}$ upper triangular and $\mathbf{N}_K$ strictly upper triangular, the BLS has an exact finite Volterra series. The necessary and sufficient condition that BLS (5.61) can be transformed to such triangular condition is computed in [D6] as follows:

$$\begin{bmatrix} \mathbf{N}_k \\ \mathbf{c}_k^T \end{bmatrix} \mathbf{A}^j [\mathbf{N}_i \mathbf{b}_i] = \mathbf{0}$$

for all $k > i$ and all $0 \leq j \leq n$. Here $\mathbf{c}_K^T$ and $\mathbf{b}_i$ are the kth row and ith column of $\mathbf{C}^T$ and $\mathbf{B}$, respectively.

A most comprehensive study of BLS realization for scalar input–scalar output with likely extensions to multidimensional cases is presented in [D2]. The realization theory of BLS is developed from the description of the system provided by the series expansion of the zero-state response. The necessary and sufficient condition that a given sequence of symmetrical Volterra kernels $\{h_i\}$ has a BLS realization is that:

1. h_1 has a proper rational Laplace transform.
2. $\{h_i\}_2^\infty$ is factorizable in the form

$$h_i(t_1, \ldots, t_i) = H(t_1)F(t_2 - t_1) \cdots F(t_{i-1} - t_{i-2})G(t_i - t_{i-1})$$

and F, G, and H have proper rational Laplace transforms.

As an important result toward the realization procedures, it is shown that a factorizable sequence of kernels is uniquely determined by its finite subsequence [D2]. Explicit existence criteria and an algorithm for finding a minimal bilinear realization of a prescribed nonlinear input–output map are given in [I1]. Walsh functions have convenient properties of orthonormality. Also, products and integrals of functions result in summations of Walsh functions. These are exploited conveniently for BLS parameter estimation from input–output data as shown for the simplified model for the generation of antibodies in Example 5.6.

EXERCISES

5.1. A common epidemic model is given by

$$\dot{x} = -\alpha u x,$$

where $u = n - x(t)$. $x(t)$ represents the population which is susceptible to a certain infection such that the sum of $x(t)$ and the infective population $u(t)$ is equal to a constant n. α is a constant.

(a) Compute $x(t)$ for a given n and $x(0) = x_0$.

(b) Compute the Volterra kernels.

5.2. Consider the second-order BLS given by

$$\dot{x}_1 = u, \qquad \dot{x}_2 = u x_1,$$

where $x_1(0) = x_2(0) = 0$.

(a) Compute the solution $x_2(t)$ in terms of $u(t)$.

(b) Compute the Volterra kernels that relate $x_1(t)$ and $x_2(t)$ to $u(t)$.

(c) Are the kernels unique? Explain.

5.3. Consider a nonlinear assembly of linear components such that $x_1 = u * h_1(t)$, $x_2 = (ux_1) * h_2(t)$, $y = (ux_2) * h_3(t)$, where $*$ denotes convolution.

(a) Compute $y(t)$ for $y(0) = 0$ in terms of $h_1(\cdot)$, $h_2(\cdot)$, $h_3(\cdot)$, and $u(t)$.

(b) Compute the Volterra kernels.

(c) Are the kernels stationary? Explain.

5.4. **(a)** Show that symmetric kernels yield the same output $\mathbf{y}(t)$ for a specified $\mathbf{u}(t)$ as the directly computed kernels.

(b) Show the general form of Volterra series solution to $y(t)$ in terms of triangular kernels, $h_{\text{tri}}(\cdot)$ such that $h(t, \sigma_1, \ldots, \sigma_n) = 0$ if $\sigma_{i+j} > \sigma_j$ for i, j positive integers. Note the upper limits of integration.

(c) Is the ordering of σ_i significant in part (b)?

5.5. If $|u(t)| \leq U$ and $\int_{-\infty}^{\infty} \cdots \int |h_k(\sigma_1, \ldots, \sigma_k)| \, d\sigma_1 \cdots d\sigma_k \leq c_k$ for all t, show that $|y(t)| \leq \sum_{k=1}^{\infty} c_k U^k$ for a scalar Volterra system.

5.6. Consider a dc motor with constant armature voltage V_a, viscous load damping D, and control field current $u(t)$. The armature current $x_1(t)$ and motor speed $x_2(t)$ are given by

$$\dot{x}_1 = -a_1 x_1 - n_1 u x_2 + \alpha,$$

$$\dot{x}_2 = -a_2 x_2 + n_2 u x_1,$$

$$y = x_1,$$

where a_1 is the ratio of armature resistance to armature inductance, L_a. n_1 is inversely proportional to L_a; a_2 is the ratio of D to J, motor-load inertia; n_2 is inversely proportional to J; and α is the ratio V_a/L_a.

(a) Compute $\mathbf{x}_h(t)$ the solution with $\mathbf{x}(0) = \mathbf{0}$ and $u(t) = 0$.

(b) Let $y(t) = x_2(t) - x_{2h}(t)$, and compute the first three triangular kernels with $x_1(0) = 0 = x_2(0)$.

5.7. Compute the inverse Laplace transform of

$$H(s_1, s_2) = \frac{1}{s_1 s_2 (s_1 + s_2)}.$$

5.8. **(a)** Compute and sketch the first four Walsh functions and the corresponding Rademacher functions from which they may be generated.

(b) Discuss the advantages and disadvantages of Walsh functions compared with Volterra series.

5.9. **(a)** Transform the following into a BLS:

$$\dot{x}_1 = a \cos x_3,$$

$$\dot{x}_2 = b \sin x_3,$$

$$\dot{x}_3 = u,$$

where a and b are constants.

(b) Compute the Volterra series for this system.

(c) Compute $x_1(t)$ for $u(t) = \delta_{-1}(t)$, the unit-step function, with $\mathbf{x}(0) = 0$.

5.10. Consider the following sixth-order system:

$$\dot{x}_1 = x_2, \qquad \dot{x}_3 = x_4, \qquad \dot{x}_5 = x_6,$$

$$\dot{x}_2 = -x_1 - 2x_2 + u,$$

$$\dot{x}_4 = -x_3 - x_4 + 2u,$$

$$\dot{x}_6 = -2x_5 - 3x_6 + x_1 x_3.$$

(a) Compute the Volterra series for $y(t) = x_5(t)$.
(b) Compute the Laplace transforms of the kernels.
(c) Compute $y(t)$ if $u(t) = \delta(t)$, the unit delta function, and all $x_i(0) = 0$, $i = 1, \ldots, 5$, except $x_6(0) = 1$.

5.11. Consider the following system, where $y(t) = x(t)$:

$$\dot{x} = \alpha(t)u(t)x$$

with $t \in [0, 1)$, $x(0) = 1$ and a control policy $u(t) = \frac{1}{2}$ for $t \in [0, \frac{1}{2})$ and $u(t) = 1$ for $t \in [\frac{1}{2}, 1)$.
(a) For $\alpha(t) = -4$, derive $y(t)$ by direct integration and compute the Walsh series expansion for $u(t)$ and $y(t)$ approximated to the fourth term.
(b) Compute $\mathbf{P}_{(2)}$, $\mathbf{P}_{(4)}$ and $\mathbf{\Lambda}_i^{(2)}$, $\mathbf{\Lambda}_i^{(4)}$ from (5.88) and (5.97).
(c) Simplify the identification equation (5.98) for the identification of $\alpha(t) = \mathbf{G}_{(q)}^T \mathbf{\Psi}_{(q)}(t)$. Write this equation under the form $\mathbf{B}_{(q)}^T = \mathbf{G}_{(q)}^T \mathbf{A}_{(q)}$ for $q = 2$ and $q = 4$ and samples at times $(2i - 1)/2q$.
(d) Identify $\alpha(t)$ for $q = 2$ and $q = 4$. Compare the mean-square error for $q = 2$ and $q = 4$, based on a true value $\alpha(t) = -4$.

5.12. Simplify the identification equation (5.98) **(a)** when all the parameters are constant, and **(b)** when the original equation is scalar with constant coefficients.

5.13. Let $\dot{x} = -4u(t)x$ with $u(t)$ as in Exercise 5.11 and $y(0) = x(0) = 1$.
(a) Write the identification equation to approximate the solution $y(t) = x(t)$.
(b) Compute the solution $y(t)$ from a fourth-order Walsh series approximation. Compare the results with the exact coefficients from Exercise 5.11(a).

5.14. In the epidemic model of Exercise 5.1,

$$\dot{x} = -\alpha u(t)x,$$

a control policy $u(t) = n - x(t)$ is implemented. α and n are constants, and n is the new bilinear control.
(a) Derive the corresponding identification equation from (5.98).
(b) For $x(0) = 1$, $\alpha = 1$, $n = 2$ and $q = 4$, identify the approximated solution $x(t)$ from samples at $t = (2i - 1)/2q$.
(c) Compute the exact solution $x(t)$ by indirect integration; approximate $x(t)$ by its first four Walsh coefficients, and compare these with the coefficients computed in part (b).

5.15. For Exercise 5.10, with the same output, input, and initial conditions, for $t \in [0, 1)$ compute the approximated output up to the fourth Walsh coefficient. Proceed as follows:
(a) Compute the Walsh coefficients for the Dirac delta function. Draw the result for $q = 4$.
(b) Compute the Walsh series approximation for the independent x_1, x_2 and x_3, x_4 blocks. Note that the initial condition, $x_1(0) = x_2(0) = x_3(0) = x_4(0) = 0$ allows us to compute analytically the solution, for any q.

(c) For $q = 4$, compute the Walsh series for x_1, x_2, x_3, x_4 and then for the integrated quadratic term $\int_0^t x_1 x_3 \, d\tau$.

(d) Write the identification equation for the system x_5, x_6 and solve for $q = 4$ and samples at times $t = (2i - 1)/2q$.

(e) Compare and plot the approximated $y(t) = x_5(t)$ obtained by the Walsh series and from the second-order Volterra series approximation of Exercise 5.10.

5.16. Consider the homogenous bilinear system

$$\dot{x}_1 = -x_1 + (x_1 + 2x_2)u,$$

$$\dot{x}_2 = x_1 + x_2 - (0.5x_1 + x_2)u,$$

where $u(t) = e^{-2t}$, and at $t = 0$, $[x_1 \; x_2] = [x_{10} \; x_{20}] = \mathbf{x}^{\mathrm{T}}(0)$. Derive the solution to the bilinear system as a function of x_{10} and x_{20}.

5.17. For the Exercise system of 5.16, where $[x_{10} \; x_{20}] = [1 \; 0]$, compute the solution $x_1(t)$ and $x_2(t)$ for $t \in [0, 4]$. Compute and compare the approximated solution as given by Picard's method.

CHAPTER 6

OPTIMAL CONTROL

6.1 INTRODUCTION

Various methodologies have been presented throughout the book for the analysis and general understanding of nonlinear system dynamics. There has been some emphasis on stability, controllability, approximate solutions, and the analysis of certain forms of nonlinear processes. Nondynamical nonlinear characteristics are analyzed conveniently in the frequency domain by means of Popov and describing-function methods. The stability criteria that are developed in this regard may be integrated directly with classical linear-regulator design (see Vol. 1). For purposes of analysis, a multidimensional, frequency-domain approach was introduced in Chapter 5 in conjunction with the Volterra series for nonlinear systems. Bilinear systems, which were shown to be quite prevalent in nature, depict the general characteristics of dynamical nonlinear processes by using many familiar properties of linear systems.

It is the purpose of this chapter to study the control system design problem with consideration of other performance criteria in addition to stability and controllability. In this respect, optimal control and adaptive control are introduced. Several particular case studies are considered in Chapter 7 and finally, suggestions are made for future directions. Unfortunately, space and time does not permit a detailed study of these topics. Only a brief sample is presented with the intent of motivation for further research and design application.

6.2 OPTIMAL CONTROL THEORY

Only a brief summary of optimal control and in particular the maximum principle is presented here. The reader is referred to any of the many excellent texts that are available on this subject (e.g., [A3], [A4], [B13], [K11], [L7], and [P4]).

Again, consider the nonlinear system

$$\frac{d\mathbf{x}}{dt} = \mathbf{f}(\mathbf{x}, \mathbf{u}), \tag{6.1}$$

where $\mathbf{x} \in \mathbb{R}^n$, $\mathbf{u} \in \mathbb{R}^m$, $\mathbf{f}(\cdot)$, and $\partial\mathbf{f}/\partial\mathbf{x}$ are continuous functions in $\mathbf{x}$ and $\mathbf{u}$. Also, it is assumed that the admissable control $\mathbf{u}(t)$ is piecewise continuous and belongs to a closed bounded set U. Now it is desired to compute the optimal control that controls (6.1) from some initial state $\mathbf{x}(t_0) = \mathbf{x}_0$ to a terminal state $\mathbf{x}(t_f) = \mathbf{x}_f$ at $t = t_f$, the terminal time. Optimality is determined by that policy which minimizes the performance index

$$J = \int_{t_0}^{t_f} f_0(\mathbf{x}, \mathbf{u})\, dt, \tag{6.2}$$

where $f_0(\cdot)$ and $\partial f_0/\partial\mathbf{x}$ are continuous functions of $\mathbf{x}$ and $\mathbf{u}$.

Numerous techniques, which stem primarily from the calculus of variations, have been derived for the solution of this problem. Dynamic programming utilizes the principle of optimality and the Bellman–Hamilton–Jacobi partial differential equation for its solution, while Pontryagin's maximum principle uses the solution to Hamilton's equations. Numerous computational algorithms have been developed, which for the most part are based on these methods. Indeed, this was a most active area of research in the 1960s and 1970s. Dynamic programming usually has the advantage of directly yielding a feedback synthesis that is computationally convenient. The maximum principle, on the other hand, often provides the form of control most directly. The latter seems most expedient for our brief presentation.

6.2.1 The Maximum Principle

To formulate the maximum principle, addend the cost function $f_0(\mathbf{x}, \mathbf{u})$ in (6.2) to the state in (6.1) so that

$$\frac{d\tilde{\mathbf{x}}}{dt} = \tilde{\mathbf{f}}(\mathbf{x}, \mathbf{u}), \tag{6.3}$$

where

$$\tilde{\mathbf{x}} = \begin{bmatrix} x_0 \\ x_1 \\ \cdot \\ \cdot \\ \cdot \\ x_n \end{bmatrix}, \qquad \tilde{\mathbf{f}} = \begin{bmatrix} f_0(\cdot) \\ f_1(\cdot) \\ \cdot \\ \cdot \\ \cdot \\ f_n(\cdot) \end{bmatrix}, \qquad \tilde{\mathbf{x}}(t_0) = \begin{bmatrix} 0 \\ x_{10} \\ \cdot \\ \cdot \\ \cdot \\ x_{n0} \end{bmatrix}$$

and $x_0(t_f) = J$ is to be minimized [P4]. Then define a scalar function

$$H(\mathbf{x}, \tilde{\mathbf{p}}, \mathbf{u}) = \tilde{\mathbf{p}}^{\mathrm{T}}\tilde{\mathbf{f}}(\mathbf{x}, \mathbf{u}), \tag{6.4}$$

where

$$\frac{d\tilde{\mathbf{p}}}{dt} = -\frac{\partial H}{\partial \tilde{\mathbf{x}}}. \tag{6.5}$$

[Note that if (6.3) is linear, (6.5) defines its adjoint system.]

Then it can be shown that the optimal control, $\mathbf{u}(t)$, which makes (6.2) a minimum, must satisfy the following conditions:

1. $\mathbf{u}(t) \in U$ maximizes (6.4), and
2. there exists a continuous nonzero $\tilde{\mathbf{p}}(t)$ which satisfies (6.5).

In summary, if $\mathbf{u}(t)$ controls system (6.1) from $\mathbf{x}(0) = \mathbf{x}_0$ to $\mathbf{x}(t_f) = \mathbf{x}_f$ so as to minimize (6.2), it is necessary that there exists a continuous nonzero costate vector $\tilde{\mathbf{p}}(t) \in \mathbb{R}^{n+1}$ which satisfies (6.5) such that $\mathbf{u}^*(t)$ maximizes (6.4), that is,

$$H(\mathbf{x}, \tilde{\mathbf{p}}, \mathbf{u}^*) = \sup_{\mathbf{u} \in U} H(\mathbf{x}, \tilde{\mathbf{p}}, \mathbf{u}). \tag{6.6}$$

For normal problems, p_0 is a nonpositive constant that may be set equal to -1 (or $+1$ as conventional for the calculus of variations and the minimum principle). Now the costate dynamics in (6.5) simplify to

$$\frac{d\mathbf{p}}{dt} = -\frac{\partial H}{\partial \mathbf{x}}, \tag{6.7}$$

where $\mathbf{p} \in \mathbb{R}^n$, and $H(\mathbf{x}, \tilde{\mathbf{p}}, \mathbf{u}) = H(\mathbf{x}, \mathbf{p}, \mathbf{u})$.

Being derived from first variations, the maximum principle only provides necessary conditions for an optimal solution. Such solutions are called extremal solutions. However, if the maximum principle, along with its corresponding transversality conditions, has a unique solution, then indeed such a solution $\mathbf{u}^*(t)$ is the unique optimal control.

It is assumed here that the initial state $\mathbf{x}_0$ and the initial time t_0 are fixed. Then it can be seen that for a fixed terminal state $\mathbf{x}_f$ at fixed time t_f, (6.1) and (6.7) have $2(n + 1)$ boundary conditions (including state, costate, and time). Consequently, there are sufficient conditions, in principle at least, to solve the resulting two-point boundary-value problem. If the terminal time is free, however, another condition must be provided. This condition may be obtained from first variations at the terminal point which show that the Hamiltonian, (6.4) with (6.5), must be zero. If the terminal state is constrained only to some manifold of points, $\theta(\mathbf{x})$, the first variation requires that the costate vector $\mathbf{p}(t)$ must be orthogonal to this terminal set $\theta(\mathbf{x})$ at the terminal time t_f (i.e., the transversality condition). If the terminal state is free, $\mathbf{p}(t_f) = \mathbf{0}$.

From (6.1) and (6.4), it is seen that $\dot{\mathbf{x}} = \partial H/\partial\tilde{\mathbf{p}}$, or

$$\frac{d\mathbf{x}}{dt} = -\frac{\partial H}{\partial \mathbf{p}}. \tag{6.8}$$

Equations (6.7) and (6.8) are called Hamilton's equations. In the classical calculus of variations, $\mathbf{p}$ is a Lagrange-multiplier vector which adjoins the system equations (6.1) to J, the performance index (6.2). It is related to dynamic programming and the Bellman–Hamilton–Jacobi equation as the negative gradient of the performance index in state space (i.e., $\mathbf{p} = -\partial J/\partial\mathbf{x}$).

From (6.4), (6.5), and (6.8), it is seen that for time-invariant systems and time-invariant cost [i.e., $\mathbf{f}(\mathbf{x}, \mathbf{u})$, $f_0(\mathbf{x}, u)$],

$$\frac{dH}{dt} = \left[\frac{\partial H}{\partial \mathbf{x}}\right]^{\mathrm{T}} \dot{\mathbf{x}} + \left[\frac{\partial H}{\partial \mathbf{p}}\right]^{\mathrm{T}} \dot{\mathbf{p}}$$

or

$$\frac{dH}{dt} = 0$$

and

$$H(\mathbf{x}, \tilde{\mathbf{p}}, \mathbf{u}) = H, \tag{6.9}$$

a constant. Furthermore, for free terminal time,

$$H(\mathbf{x}, \tilde{\mathbf{p}}, \mathbf{u}) = 0. \tag{6.10}$$

For time-optimal control, $J = t_f$, $f_0 = 1$, and the terminal time is free. Then the maximum principle simplifies somewhat. In this case let

$$\mathcal{H}(\mathbf{x}, \mathbf{p}, \mathbf{u}) = \mathbf{p}^{\mathrm{T}}\mathbf{f}(\mathbf{x}, \mathbf{u}), \tag{6.11}$$

where $\mathbf{p} \in \mathbb{R}^n$. Then $H(\cdot) = \mathcal{H}(\cdot) + p_0 f_0$ and the optimal control $\mathbf{u}^*$ must maximize $H(\cdot)$, or

$$\mathcal{H}(\mathbf{x}, \mathbf{p}, \mathbf{u}^*) = \sup_{\mathbf{u} \in U} \mathcal{H}(\mathbf{x}, \mathbf{p}, \mathbf{u}). \tag{6.12}$$

Then the Hamiltonian is a positive constant (for the normal case with $p_0 < 0$),

$$\mathcal{H} > 0 \tag{6.13}$$

and the costate of (6.8) is given by

$$\frac{d\mathbf{p}}{dt} = -\frac{\partial \mathcal{H}}{\partial \mathbf{x}}. \tag{6.14}$$

Example 6.1

Consider a simple first-order bilinear system, which also represents the prompt neutron kinetics [i.e., (2.122) in Vol. 1 without precursors for fast transients],

$$\dot{x} = ux, \tag{6.15}$$

where $|u| \leq 1$. Suppose that it is desired to control the process from $x(0) = x_0$ to $x(t_f) = x_f$ in minimum time.

From (6.11) and (6.14),

$$\mathcal{H} = uxp,$$

$$\dot{p} = -up.$$

Then from (6.12), the extremal policy is bang-bang such that

$$u = \operatorname{sgn} p(t)x(t),$$

where $x(t)$ is positive and

$$p = p_0 \exp \int_0^t -u \, d\sigma.$$

Since $\mathcal{H}$ is a positive constant with $p(t)$ continuous and

$$x = x_0 \exp \int_0^t u \, d\sigma,$$

$u(t)$ cannot change sign. Therefore,

$$u(t) = \begin{cases} +1 & \text{for } x_f > x_0, \\ -1 & \text{for } x_f < x_0, \end{cases}$$

where $t \in (0, t_f)$. If it is desired that $x(t) = x_f$ for $t \geq t_f$, then $u(t) = 0$ for $t > t_f$. It is readily apparent that the solution is unique, and therefore the extremal $u(t)$ is indeed optimal.

Example 6.2

Consider the scalar bilinear regulator

$$\dot{x} = ax + nux, \tag{6.16}$$

where a and n are constants and n is positive. Here it is desired to control the BLS (6.16) from some $x(0) = x_0$ to 0 so as to minimize

$$J = \tfrac{1}{2} \int_0^\infty (qx^2 + ru^2)\, dt, \tag{6.17}$$

where q and r are positive constants. Relative to the formulation above, the infinite horizon control is assumed such that

$$\lim_{t \to \infty} x(t) = 0.$$

Otherwise, since (6.16) is not controllable to the origin with finite control, the problem would not be well posed.

From (6.4) and (6.5), with $p_0 = -1$,

$$H(x, p, u) = -\tfrac{1}{2}(qx^2 + ru^2) + apx + npux \tag{6.18}$$

and

$$\dot{p} = qx - ap - nup. \tag{6.19}$$

Here u is constrained in terms of weighting "control energy" relative to mean-square state error in the performance index (6.17). Then it is necessary for $H(\cdot)$ to be a maximum that

$$\frac{\partial H}{\partial u} = -ru + npx = 0 \tag{6.20}$$

or that

$$u = \frac{n}{r} p(t) x(t). \tag{6.21}$$

It is desired to express $u(p, x)$ as a function $u(x)$ that would result in a nonlinear feedback control. Consequently, let $p(t) = F(x)$, so that

$$\dot{p} = \frac{dF(x)}{dx}\frac{dx}{dt} = \frac{dF(x)}{dx}(ax + nux),$$

and from (6.19),

$$\dot{p} = qx - F(x)(a + nu).$$

Then with $F_1(x) = xF(x)$,

$$\frac{dF_1(x)}{dx} = \frac{qx}{a + nu}.$$

Then, from (6.21),

$$\frac{dF_1(x)}{dx} = \frac{qx}{a + n^2 F_1(x)/r}.$$

Separation of variables and direct integration yield

$$aF_1(x) + \frac{n^2}{r[F_1^2(x)/2]} = \frac{qx^2}{2} + \frac{c}{2},$$

where c is a constant of integration. Then

$$F_1(x) = \tfrac{1}{2}w \pm \tfrac{1}{2}[w^2 + 4z(x)]^{1/2}, \tag{6.22}$$

where $w = -2ar/n^2$ and $z(\cdot) = r(qx^2 + c)/n^2$. Substitution of (6.22) into (6.21) into (6.16) shows that

$$\dot{x} = \pm\left(a^2 + \frac{cn^2}{r} + \frac{qn^2x^2}{r}\right)^{1/2} x. \tag{6.23}$$

Let $\alpha = (a^2 + cn^2/r)^{1/2}$. Then with $x(0) = x_0$ and $x(\infty) = 0$, it can be shown that the solution to (6.23) requires that $\alpha = 0$, or $c = -a^2r/n^2$.

The closed-loop system (6.23) is described by

$$\dot{x} = \pm n\left(\frac{q}{r}\right)^{1/2} x^2$$

or

$$x(t) = \frac{x_0}{1 \pm x_0 n(q/r)^{1/2} t}. \tag{6.24}$$

The optimal control (6.21) becomes

$$u(t) = \frac{-a}{n} \pm \left(\frac{q}{r}\right)^{1/2} x. \tag{6.25}$$

From continuity of $x(t)$ and $p(t)$ it becomes apparent that the optimal control is given by (6.25) with the $-$ sign when x_0 and n are of same sign; otherwise, the $+$ is used. In other words,

$$u = -\frac{a}{n} + [\text{sgn}\,(nx_0)]\left(\frac{q}{r}\right)^{1/2} x \tag{6.26}$$

and

$$x(t) = \frac{x_0}{1 + |nx_0|(q/r)^{1/2} t}. \tag{6.27}$$

Suppose the problem is changed so that $x(t_f) = \xi$, where ξ is small and

$$J = \frac{1}{2}\int_0^{t_f} (qx^2 + ru^2)\, dt. \tag{6.28}$$

Now the finite terminal state can be reached in finite time, and the problem is well posed. Here t_f is free, so that from (6.18),

$$H = -\tfrac{1}{2}(qx^2 + ru^2) + apx + npux = 0.$$

Then substitution of u from (6.21) yields

$$H = -\frac{qx^2}{2} + \frac{n^2p^2x^2}{2r} + apx = 0. \tag{6.29}$$

In this case (6.29) shows that

$$p = F(x) = -\frac{ar}{n^2x}\left[1 \pm \left(\frac{1 + qn^2x^2}{a^2r}\right)^{1/2}\right] \tag{6.30}$$

and from (6.21) with $\beta = (q/r)(n/a)^2$,

$$u = -\frac{a}{n}[1 \pm (1 + \beta x^2)^{1/2}]. \tag{6.31}$$

Then (6.16), the closed-loop system model, becomes

$$\dot{x} = \pm a[1 + \beta x^2]^{1/2} x, \tag{6.32}$$

where the $-$ sign is used if $a > 0$ and $+$ sign if $a < 0$. Note that (6.32) is identical to (6.23) if $c = 0$ as (6.29) requires.

If control energy and mean-square error are equally weighted, $q = r = 1$, and

$$u = -\frac{a}{n} \mp \left(\frac{a^2}{n^2} + x^2\right)^{1/2}.$$

As $r \to 0$, the state feedback gain approaches infinity. As $q \to 0$, it approaches zero.

In case $|u| \leq 1$ and $r = 0$, then from (6.18) the optimal control takes the form

$$u = \operatorname{sgn} px.$$

Consequently, the maximum principle is satisfied by

$$u = -1$$

for $t \in (0, t_f)$.

In all cases with finite t_f, $\dot{x}(t) = 0$ with $x(t) = \xi$, and $t > t_f$ if $u(t) = -a/n$. Of course, $|a/n| \leq 1$ is required for this terminal control to be admissable in the latter case above. Again, the solutions to the maximum principle are unique for this problem and the control is optimal.

6.2.2 Quadratic Performance Index

The scalar bilinear regulator of Example 6.2 may be generalized somewhat according to [T1]. This nice result is briefly summarized here without proof. Consider the scalar-controlled bilinear system

$$\frac{d\mathbf{x}}{dt} = \mathbf{A}\mathbf{x} + \mathbf{N}u\mathbf{x} + \mathbf{b}u, \tag{6.33}$$

where $\mathbf{x} \in \mathbb{R}^n$, and $\mathbf{A}$, $\mathbf{N}$, $\mathbf{b}$ are constant and of appropriate dimensions. Let the performance index be given by

$$J = \tfrac{1}{2}\int_{t_0}^{t_f} [\mathbf{x}^{\mathrm{T}}\mathbf{Q}_1(\mathbf{x}, t)\mathbf{x} + r(t)u^2]\, dt + \tfrac{1}{2}\mathbf{x}^{\mathrm{T}}(t_f)\mathbf{P}_f\mathbf{x}(t_f), \tag{6.34}$$

where $\mathbf{Q}_1(\cdot)$ and $\mathbf{P}_f$ are symmetric, positive-definite matrices, and $r(\cdot)$ is a positive scalar function for $t \in [t_0, t_f]$. It is assumed that the initial state $\mathbf{x}(t_0) = \mathbf{x}_0$ is fixed, and the terminal state and time are free. Then it is shown in [T1] that the following quadratic control is stabilizing and makes (6.34) a minimum for BLS (6.33):

$$u(t) = -r^{-1}(t)\mathbf{S}^{\mathrm{T}}(\mathbf{x})\mathbf{P}(t)\mathbf{x}(t), \tag{6.35}$$

where $\mathbf{S}(\mathbf{x}) = \mathbf{b} + \mathbf{N}\mathbf{x}$,

$$\dot{\mathbf{P}} = -\mathbf{A}^{\mathrm{T}}\mathbf{P} - \mathbf{P}\mathbf{A} + \mathbf{P}\mathbf{b}r^{-1}\mathbf{b}^{\mathrm{T}}\mathbf{P} - \mathbf{Q}_2,$$

$$\mathbf{P}(t_f) = \mathbf{P}_f, \qquad \mathbf{Q}_2(t) > 0 \qquad \text{(i.e., positive definite)},$$

$$\mathbf{Q}_2(t) - \mathbf{P}(t)\mathbf{b}r^{-1}\mathbf{b}^{\mathrm{T}}(t) > 0,$$

and $\mathbf{Q}_1(\mathbf{x}, t)$ in the performance index has a special form

$$\mathbf{Q}_1(\mathbf{x}, t) = \mathbf{Q}_2(t) + \mathbf{P}(t)[\mathbf{s}(\mathbf{x})r^{-1}\mathbf{s}^{\mathrm{T}}(\mathbf{x}) - \mathbf{b}r^{-1}\mathbf{b}^{\mathrm{T}}]\mathbf{P}(t).$$

However, note that the general bilinear regulator is not solved, and only the special form of $\mathbf{Q}_1(\mathbf{x}, t)$ permits a solution here. The solution is related to the maximum principle of Section 6.2.1 with costate $\mathbf{p}(t) = -\mathbf{P}(t)\mathbf{x}(t)$. As a result of the terminal cost function $T = \frac{1}{2}\mathbf{x}^{\mathrm{T}}(t_f)\mathbf{P}_f\mathbf{x}(t_f)$ the terminal costate $\mathbf{p}(t_f) = -\partial \mathrm{T}/\partial \mathbf{x} = -\mathbf{P}_f\mathbf{x}(t_f)$ instead of $\mathbf{p}(t_f) = \mathbf{0}$ for the free-terminal-state problem with no terminal cost as given in Section 6.2.1.

A more conventional quadratic performance index is considered for this problem in [H2]. Here the solution is approximated by a convergent sequence of linear regulator solutions. Again, the system is given by (6.33) with a quadratic performance index

$$J = \frac{1}{2}\int_{t_0}^{t_f} (\mathbf{x}^{\mathrm{T}}\mathbf{Q}\mathbf{x} + ru^2)\,dt + \frac{1}{2}\mathbf{x}^{\mathrm{T}}(t_f)\mathbf{P}_f\mathbf{x}(t_f), \tag{6.36}$$

where $\mathbf{Q}$ and $\mathbf{P}_f$ are constant symmetric positive-definite matrices, and r is a positive constant scalar. Again, the initial state $\mathbf{x}(0) = \mathbf{x}_0$ is fixed, but the terminal conditions are free.

From (6.4) with $p_0 = -1$, the Hamiltonian becomes

$$H(\mathbf{x}, u, \mathbf{p}) = -\frac{1}{2}(\mathbf{x}^{\mathrm{T}}\mathbf{Q}\mathbf{x} + ru^2) + \mathbf{p}^{\mathrm{T}}(\mathbf{A}\mathbf{x} + \mathbf{N}u\mathbf{x} + \mathbf{b}u), \tag{6.37}$$

with maximizing control, from $\partial H/\partial u = 0$,

$$u = \frac{(\mathbf{N}\mathbf{x} + \mathbf{b})^{\mathrm{T}}\mathbf{p}}{r}, \tag{6.38}$$

where $\mathbf{p}(t)$ satisfies the costate equation (6.5). With the terminal cost function included in (6.36), it can be seen from the above or from the first variations at the terminal end that

$$\mathbf{p}(t_f) = -\mathbf{P}_f\mathbf{x}(t_f). \tag{6.39}$$

From (6.33) and (6.5), the two-point boundary-value problem is specified by

$$\dot{\mathbf{x}} = \mathbf{A}\mathbf{x} + (\mathbf{N}\mathbf{x} + \mathbf{b})r^{-1}(\mathbf{N}\mathbf{x} + \mathbf{b})^{\mathrm{T}}\mathbf{p}, \tag{6.40}$$

$$\dot{\mathbf{p}}_i = [\mathbf{Q}\mathbf{x}]_i - [\mathbf{A}^{\mathrm{T}}\mathbf{p}]_i - \mathbf{p}^{\mathrm{T}}[\mathbf{N}_i^{\mathrm{T}}r^{-1}(\mathbf{N}\mathbf{x} + \mathbf{b})^{\mathrm{T}} + (\mathbf{N}\mathbf{x} + \mathbf{b})r^{-1}\mathbf{N}_i]\mathbf{p},$$

where $[\,\cdot\,]_i$ designates the ith row, $\mathbf{x}(0) = \mathbf{x}_0$, and $\mathbf{p}(t_f) = -\mathbf{P}_f\mathbf{x}(t_f)$.

To formulate a sequence of linear system solutions, represent (6.40) by its corresponding time-variant linear case (i.e., $\mathbf{N} = \mathbf{0}$):

$$\dot{\mathbf{x}} = \tilde{\mathbf{A}}(t)\mathbf{x} - \tilde{\mathbf{b}}(t)r^{-1}\tilde{\mathbf{b}}(t)^{\mathrm{T}}\mathbf{p},$$

$$\dot{\mathbf{p}} = \tilde{\mathbf{Q}}(t)\mathbf{x} - \tilde{\mathbf{A}}(t)^{\mathrm{T}}\mathbf{p}, \tag{6.41}$$

where from (6.40) the time-variant matrices may be evaluated as functions of the constant matrices in (6.40), and of $\mathbf{p}(t)$ and $\mathbf{x}(t)$.

Then the iterative solution of (6.40) proceeds with first the linear system solution

$$\dot{\mathbf{x}}^{(0)} = \mathbf{A}\mathbf{x}^{(0)} - \mathbf{b}(t)r^{-1}\mathbf{p}^{(0)}, \qquad \mathbf{x}^0(0) = \mathbf{x}_0,$$

$$\dot{\mathbf{p}}^{(0)} = \mathbf{Q}\mathbf{x}^{(0)} - \mathbf{A}^{\mathrm{T}}\mathbf{p}^{(0)}, \qquad \mathbf{p}^0(t_f) = \mathbf{P}_f\mathbf{x}^0(t_f). \tag{6.42}$$

Then, with $k = 0, 1, \ldots,$

$$\dot{\mathbf{x}}^{(k+1)} = \tilde{\mathbf{A}}^{(k)}(t)\mathbf{x}^{(k+1)} - \tilde{\mathbf{b}}^{(k)}(t)r^{-1}\tilde{\mathbf{b}}^{(k)}(t)\mathbf{p}^{(k+1)}, \qquad \mathbf{x}^{(k+1)}(0) = \mathbf{x}_0,$$

$$\dot{\mathbf{p}}^{(k+1)} = \tilde{\mathbf{Q}}^{(k)}\mathbf{x}^{(k+1)} - \tilde{\mathbf{A}}^{(k)\mathrm{T}}\mathbf{p}^{(k+1)}, \qquad \mathbf{p}^{(k+1)}(t_f) = \mathbf{P}_f\mathbf{x}^0(t_f). \tag{6.43}$$

Here $\tilde{\mathbf{A}}^{(k)} = \tilde{\mathbf{A}}^{(k)}(\mathbf{p}^{(k)}(t))$, and so on.

As for the linear-regulator problem [A3],

$$\mathbf{p}^{(k+1)}(t) = -\mathbf{K}^{(k+1)}(t)\mathbf{x}^{(k+1)}(t), \tag{6.44}$$

where $\mathbf{K}^{(k+1)}$ is the solution of a Riccati equation

$$\dot{\mathbf{K}}^{(k+1)} = -\tilde{\mathbf{Q}}^{(k)} - \mathbf{K}^{(k+1)}\tilde{\mathbf{A}}^{(k)} - [\tilde{\mathbf{A}}^{(k)\mathrm{T}}\mathbf{K}^{(k+1)}$$

$$+ \mathbf{K}^{(k+1)}\tilde{\mathbf{b}}^{(k)}r^{-1}\tilde{\mathbf{b}}^{(k)\mathrm{T}}\mathbf{K}^{(k+1)}]\mathbf{x}^{(k+1)}. \tag{6.45}$$

Here $\mathbf{x}^{(k+1)}(0) = \mathbf{x}^0$. Then the feedback controller becomes as for the linear case,

$$u^{(k+1)}(t) = r^{-1}[\mathbf{N}\mathbf{x}^{(k+1)}(t) + \mathbf{b}]^{\mathrm{T}}\mathbf{K}^{(k+1)}(t)\mathbf{x}^{(k+1)}(t). \tag{6.46}$$

It is shown in [H2] that the sequence is convergent by suitable selection of $\mathbf{Q}$, $\mathbf{P}_f$, and r.

Example 6.3 [H2]

Consider a bilinear process for a continuously stirred tank reactor with an exothermic reaction with

$$\dot{x}_1 = \frac{13}{6}x_1 + \frac{5}{8}x_2 - ux_1 - \frac{u}{8},$$

$$\dot{x}_2 = \frac{50}{3}x_1 + \frac{8}{3}x_2,$$

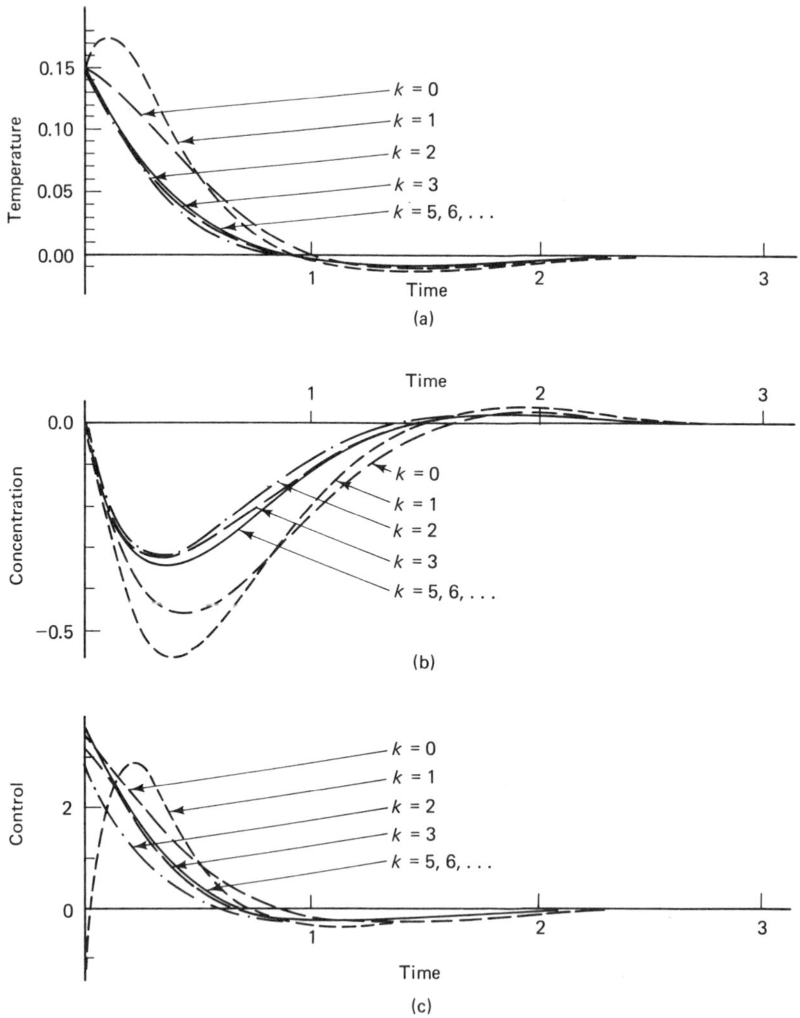

Figure 6.1 Time responses for Example 6.3. (After [H2].)

where x_1 and x_2 are initial reaction product temperature and concentration, respectively; u, the normalized dimensionless control, represents the coolant flow rate in the reactor jacket. The performance index is given by (6.36) with

$$r = 1, \qquad \mathbf{Q} = \begin{bmatrix} 10 & 0 \\ 0 & 10 \end{bmatrix}, \qquad \mathbf{P}_f = \begin{bmatrix} 1000 & 0 \\ 0 & 1000 \end{bmatrix}.$$

Time is normalized and dimensionless with $t_f = 3$. Figure 6.1 shows the sequential computer solutions for $k = 0, \ldots, 6$, which depicts satisfactory convergence with $\mathbf{x}_0^{\mathrm{T}} = [0.15 \quad 0]$ for temperature, concentration, and control.

The optimal quadratic-cost control for bilinear systems with constrained control is analyzed by adding a nonnegative state penalty to the cost function in [R6]. If quadratic control penalty is incorporated, continuous extremal controls usually result, and without control penalty bang-bang controls usually evolve with the possibility of continuous singular solutions.

Again consider BLS (6.33) with $|u| \leq 1$, t_f fixed, and

$$J = \int_{t_0}^{t_f} \{\tfrac{1}{2}[\mathbf{x}^{\mathrm{T}}\mathbf{Q}(t)\mathbf{x} + r(t)u^2] + g(\mathbf{x}, t)\}\, dt + \tfrac{1}{2}\mathbf{x}^{\mathrm{T}}\mathbf{P}_f(t_f)\mathbf{x}. \tag{6.47}$$

Here $\mathbf{Q}(\cdot)$ is symmetric and positive definite, and $r(t)$ is a positive function in one case and is zero in another case. If $r = 0$, it can be seen from the maximum principle that

$$u = -\operatorname{sgn} s(\mathbf{x}, t), \tag{6.48}$$

where including time-variant (6.33),

$$s(\mathbf{x}, t) = \{\mathbf{K}(t)[\mathbf{b}(t) + \mathbf{N}(t)\mathbf{x}]\}^{\mathrm{T}}\mathbf{x}, \tag{6.49}$$

and

$$\mathbf{K}(t) = \mathbf{\Phi}^{\mathrm{T}}(t_f, t)\mathbf{P}_f\mathbf{\Phi}(t_f, t) + \int_t^{t_f} \mathbf{\Phi}^{\mathrm{T}}(\tau, t)\mathbf{Q}(\tau)\mathbf{\Phi}(\tau, t)\, d\tau \geq 0, \tag{6.50}$$

$t \in [t_0, t_f]$; $\mathbf{\Phi}(\cdot)$ is the state-transition matrix of the linear subsystem of matrix $\mathbf{A}$. Here

$$\mathbf{K}(t_f) = \mathbf{P}_f, \tag{6.51}$$

and the solution assumes a cost-addended function

$$g(x, t) = |s(x, t)|. \tag{6.52}$$

When $\mathbf{x}(t) = \mathbf{0}$, $u(t) = 0$ is a solution. More generally, if $s(x, t) \equiv 0$ for some finite time interval with the maximum principle satisfied, the resulting control is said to be singular since (6.48) is undefined. In such a case the singular control must be considered in the optimal synthesis. The required conditions for a singular solution to this problem are somewhat complex but are derived in [R6].

If $t_f = \infty$, the policies above still satisfy the maximum principle with the optimal control a bang-bang policy with $u = 0$ if $\mathbf{x} = \mathbf{0}$, and the possibility of other singular cases. However, in this case $\mathbf{K}$ is a constant solution to Liapunov's equation,

$$\mathbf{KA} + \mathbf{A}^{\mathrm{T}}\mathbf{K} + \mathbf{Q} = \mathbf{0},$$

which yields

$$\mathbf{K} = \int_0^\infty e^{\mathbf{A}^{\mathrm{T}}t}\mathbf{Q}e^{\mathbf{A}t}\,dt. \tag{6.53}$$

If $r(t) > 0$ for all $t \in [t_0, t_f]$, it is shown in [R6] that the optimal control is given by

$$u(\mathbf{x}, t) = -\mathrm{sat}\,[r^{-1}(t)s(\mathbf{x}, t)], \tag{6.54}$$

where $s(\mathbf{x}, t)$ is given by (6.49) with $\mathbf{K}(t)$ given by (6.50). As in Vol. 1, the saturation function is defined by

$$\mathrm{sat}\,(\sigma) = \begin{cases} +1, & \sigma > 1 \\ \sigma, & |\sigma| \le 1 \\ -1, & \sigma < 1. \end{cases} \tag{6.55}$$

For this case, however, $g(\mathbf{x}, t)$ in (6.47) is given by

$$g(\mathbf{x}, t) = \left[\begin{array}{ll} |s(\mathbf{x}, t)| - \dfrac{r(t)}{2}, & |s(\mathbf{x}, t)| > r(t) \\ \frac{1}{2}r^{-1}(t)s^2(\mathbf{x}, t), & |s(\mathbf{x}, t)| \le r(t). \end{array}\right. \tag{6.56}$$

If $\mathbf{A}$ is Hurwitz and r is a positive constant, the infinite horizon optimal control (i.e., the above with $t_f = \infty$) is optimal and the origin is asymptotically stable. Again, however, $\mathbf{K}$ in (6.49) is a constant matrix given by (6.53).

Example 6.4 [R6]

Consider

$$\begin{aligned} \dot{x}_1 &= x_2, \\ \dot{x}_2 &= -x_1 - \alpha x_2 - \alpha u x_2, \end{aligned} \tag{6.57}$$

where $|u| \leq 1$, $\alpha > 0$. Assume that in the performance index (6.47) $\mathbf{Q} = 2\mathbf{I}$. Then (6.53) yields

$$\mathbf{K} = \begin{bmatrix} \alpha + 2\alpha^{-1} & 1 \\ 1 & 2\alpha^{-1} \end{bmatrix},$$

and from (6.49),

$$s(\mathbf{x}) = (\mathbf{KNx})^{\mathrm{T}}\mathbf{x} = -x_2(\alpha x_1 + 2x_2). \tag{6.58}$$

But $s(\mathbf{x}) = 0$ if $x_1 = x_2 = 0$ or if $x_2 = -\alpha x_1/2$. The latter is satisfied for the system above if $u = 2\alpha^{-2} - \alpha^{-1} - 1$, and the former if $u = 0$.

In summary, the optimal control,

$$u(\mathbf{x}) = \begin{cases} \operatorname{sgn} s(x) & \text{if } s(x) \neq 0 \\ 2\alpha^{-2} - \alpha^{-1} - 1 & \alpha x_1 + 2x_2 = 0 \\ 0 & \mathbf{x} = \mathbf{0}, \end{cases} \tag{6.59}$$

minimizes

$$J = \int_0^{\infty} [(x_1^2 + x_2^2) + |s(\mathbf{x})|]\, dt. \tag{6.60}$$

The optimal trajectories are shown on Fig. 6.2.

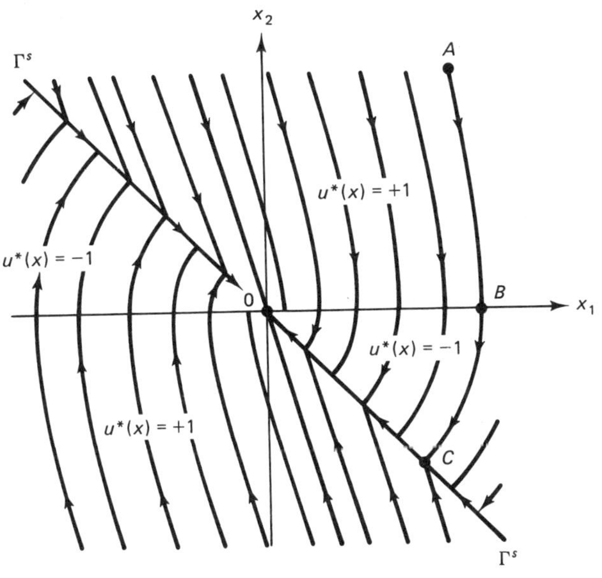

Figure 6.2 Optimal bang-bang feedback control; Example 6.4. (After [R6].)

If $r > 0$, from (6.54),

$$u(\mathbf{x}) = -\text{sat}\,[r^{-1}s(\mathbf{x})], \tag{6.61}$$

where $s(\mathbf{x})$ is given by (6.58). Here (6.61) minimizes (6.47) with $\mathbf{Q} = \mathbf{I}$, and $g(\cdot)$ given by (6.56).

6.3 VARIABLE-STRUCTURE PROCESSES

Switching types of controllers are considered in this section. The emphasis here is on relatively simple nonlinear systems of variable structure such as the bilinear system and more general systems that are linear in control but nonlinear in state. Optimal regulation, time-optimal control, and fuel-optimal control problems with control constraints, and in some cases, state constraints, will be studied. Although such constraints are in some sense taken into account with a quadratic cost in state and control, there frequently exist certain fixed physical limitations. For example, the valve can only open so far, the aileron can only turn so far, and the reactor can only get so hot. The optimal control study relative to such constraints can also provide valuable information for the system designer relative to critical design parameters. For example, the minimum coolant flow rate of a rocket engine or jet engine, used during startup to maximize payload, may be constrained by certain component temperature or by a pump stall condition (so-called "pump stall specific speed"). Such information not only may provide a high-performance optimal control, but may also lead to improved plant design. It also shows the need for integrated dynamic analysis, plant design, and control design.

6.3.1 Nonlinear Systems with Linear Control

Here we consider the general class of nonlinear systems which are linear in control with

$$\frac{d\mathbf{x}}{dt} = \mathbf{a}(\mathbf{x}) + \mathbf{b}(\mathbf{x})u, \tag{6.62}$$

where again $\mathbf{x} \in \mathbb{R}^n$; $\mathbf{a}(\cdot)$ and $\mathbf{b}(\cdot)$ are continuous n-vector functions with continuous derivatives, and u is assumed scalar for convenience. Further, assume that the cost function given by (6.2) is linear in control so that

$$f_0(\mathbf{x}, u) = a_0(\mathbf{x}) + b_0(\mathbf{x})u, \tag{6.63}$$

where $a_0(\cdot)$ and $b_0(\cdot)$ are continuous scalar functions that have continuous derivatives. Obviously, the system and its cost function are a generalization of the addended bilinear system which is linear in state as well as control.

It was shown early by Balakrishnan [B2] that under quite general conditions the dynamics of controllable nonlinear processes can be defined by (6.62). While mathematical existence and uniqueness of optimal solutions can be quite difficult, it has also been shown that under general controllability conditions, the optimal controls problem from an initial state to a compact target set for (6.62) and (6.63) do admit a solution. Consequently, optimal control problems for this general class were investigated by several investigators as demonstrated by [H1], [M19], [A4], and [K12].

The maximum principle (6.6) shows directly that the optimal control for (6.62) and (6.63) is a bang-bang process with the possibility of singular solutions such as depicted for the bilinear systems in Section 6.2.2 when the switching function is zero for a finite time interval.

In the next section, the control to a terminal set $\theta(\mathbf{x})$ is studied. Sufficient conditions of optimality for such problems are provided by [B14] as adapted in [R1] and [M2]. Essentially, the state-space region is divided into a finite number of subspaces by switching hypersurfaces $\mathcal{S}$ such that

1. $\partial u(\mathbf{x})/\partial \mathbf{x}$ is continuous on the subspaces interior to $\mathcal{S}$,
2. $\mathbf{x}(t)$ intersects $\mathcal{S}$ with nonzero angle at a finite number of points and eventually intersects $\theta(\mathbf{x})$ after elapsed time $t(\mathbf{x})$, and
3. the performance index $J(\mathbf{x})$ to reach $\theta(\mathbf{x})$ is a continuous function on all of $\mathbb{R}^n$.

6.3.2 Regulation

Consider the following second-order regulator with classical position feedback and controlled-rate feedback:

$$\begin{aligned} \dot{x}_1 &= x_2, \\ \dot{x}_2 &= -x_1 - \alpha(1 + u)x_2, \end{aligned} \tag{6.64}$$

where x_1 is a position error, α is a positive constant, and $|u| \leq 1$. The optimal control is to transfer an initial state to a small circle of radius ρ about the origin in such a manner as to minimize

$$J(\mathbf{x}) = \int_0^{t(\mathbf{x})} [\mathbf{x}_1^2(\tau) + \mathbf{x}_2^2(\tau)]\, d\tau. \tag{6.65}$$

A family of extremal trajectories can be generated by running both the state and costate systems backward in time from "initial" points $\mathbf{x}(0)$ on $\mathcal{S}$, provided that initial conditions can be established for $\mathbf{p}(0)$.

If $\mathbf{x}(t)$ is an extremal trajectory from $\mathbf{x}$ to $\mathbf{x}(0) \in \theta$, the maximum principle requires that there exist a costate vector function $\mathbf{p}(t)$ which is defined by the adjoint system of (6.64) with $\mathbf{p}(0)$ normal to θ at $\mathbf{x}(0)$. In the present case, $\mathbf{p}(0)$ must be parallel to the vector $\mathbf{x}(0)$ since θ is a circle. Thus set $\mathbf{p}(0) = \lambda\,\mathbf{x}(0)$, where λ is a negative number that remains to be specified.

From the maximum principle, (6.4) and (6.7) yield

$$H = -(x_1^2 + x_2^2) + p_1x_2 - p_2\alpha(1 + u)x_2 - p_2x_1, \tag{6.66}$$

$$\begin{aligned} \dot{p}_1 &= 2x_1 + p_2, \\ \dot{p}_2 &= 2x_2 - p_1 + p_2\alpha(1 + u). \end{aligned} \tag{6.67}$$

The value of $u(t)$ along an extremal, as deduced from the maximum of (6.66), is

$$u(t) = -\text{sgn}\,[p_2(t)x_2(t)]. \tag{6.68}$$

Suppose that there exists a singular arc, that is, a segment of extremal along which $p_2(t)x_2(t) = 0$ for $t \in [t_a, t_b]$. This implies that either $x_2(t) = \dot{x}_2(t) = 0$ or $\dot{p}_2(t) = p_2(t) = 0$ for $t \in [t_a, t_b]$. The former case is impossible since (6.64) would imply that $x_1(t) = 0$ as well, and the origin in state space is not exterior to θ. In the latter case, however, the adjoint system yields $\dot{p}_1 = 2x_1$ and $p_1 = 2x_2$. Therefore,

$$\begin{aligned} \dot{x}_2 &= x_1 = -x_1 - \alpha(1 + u)x_2, \\ x_1 &= -\tfrac{1}{2}\alpha(1 + u)x_2 \end{aligned} \tag{6.69}$$

along the singular arc. Also, substitution of $p_2 = 0$ and $p_1 = 2x_2$ into the Hamiltonian $H = 0$, where

$$H(\mathbf{x}, \mathbf{p}, \mathbf{u}) = [\mathbf{f}(\mathbf{x}, \mathbf{u})]^{\mathrm{T}}\mathbf{p} - f_0(\mathbf{x}, \mathbf{u})$$

with $p_0 = -1$ for convenience, yields the further condition

$$x_1^2 = x_2^2.$$

If $x_1 = x_2$, (6.69) implies that $u < -1$, which is not in U. Thus the only singular arcs are on the line $x_1 = -x_2$, and the control value along such an arc is

$$u = \frac{2}{\alpha} - 1,$$

which is in U for $\alpha \geq 1$.

The extremals that are generated by solving the reversed-time state and costate equations for initial states $\mathbf{x}(0)$ on θ, with u determined by (6.68), do not contain any of the singular arcs above. Also, it is shown below that these extremals satisfy sufficient conditions for optimality, and hence they are optimal. It follows that an extremal containing a singular arc has, at best, an equal cost functional.

It is necessary to evaluate λ in order to determine the initial costate vector $\mathbf{p}(0) = \lambda\mathbf{x}(0)$. Since $p_2(0) = \lambda x_2(0)$, (6.68) requires that $u(0) = +1$. Then $H = 0$ if and only if

$$\lambda = -\frac{x_1^2(0) + x_2^2(0)}{2\alpha x_2^2(0)} \tag{6.70}$$

which exists almost everywhere on $\mathcal{S}$.

The families of extremals for $\alpha = 1$ and $\alpha = \frac{1}{2}$ are shown in Figs. 6.3 and 6.4. In each case the extremals cover the entire plane and do not intersect one another, and the corresponding cost functional is continuous.

In each figure, the plane is divided by "switching curves" into regions on which u is constant. These switching curves constitute the set $\mathcal{S}$, and it is evident that the extremals cross $\mathcal{S}$ with nonzero angles. In each case the $x_2 = 0$ axis is a switching curve because of (6.68). For the case $\alpha = \frac{1}{2}$, the eigenvalues of the system with $u = +1$ are complex, and two switching curves are shown in Fig. 6.4 which have

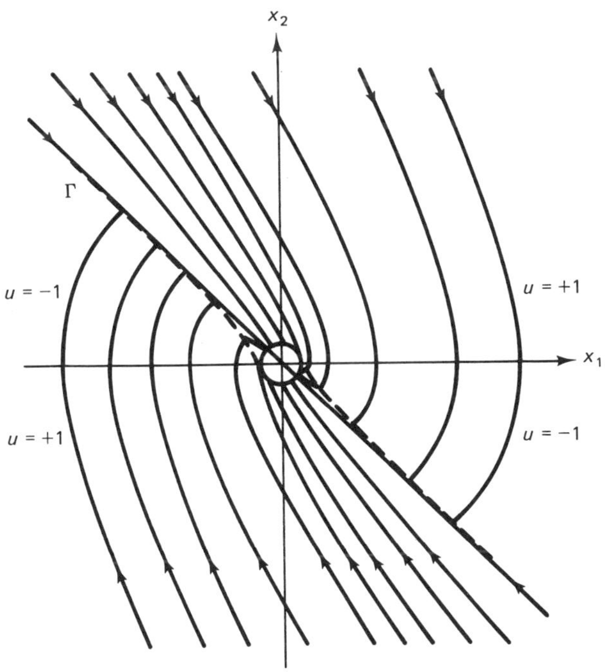

Figure 6.3 Optimal bilinear regulation, $\alpha = 1$. (From [M2].)

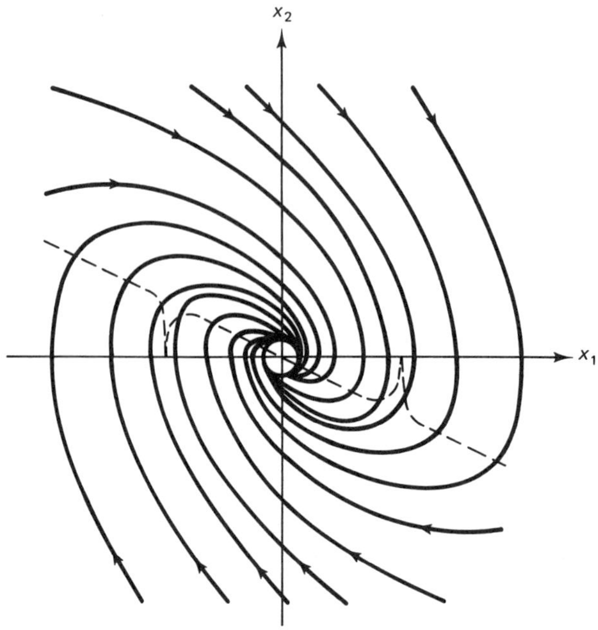

Figure 6.4 Optimal bilinear regulation, $\alpha = \frac{1}{2}$. (From [M2].)

the looped characteristic of switching curves for linear systems with complex eigenvalues. The control system state diagram is shown in Fig. 6.5.

It is of interest to examine the effect of changing the radius ρ on the extremal synthesis. Suppose that $\mathbf{x}$ is any extremal terminating on a circle θ_1 with radius ρ_1, and that $\mathbf{p}$ is the corresponding costate trajectory, with $\mathbf{p}(0) = \lambda\mathbf{x}(0)$. The control function $u(t)$ along the trajectory is given by (6.68).

Note that the state–costate system is linear and homogeneous, and that expres-

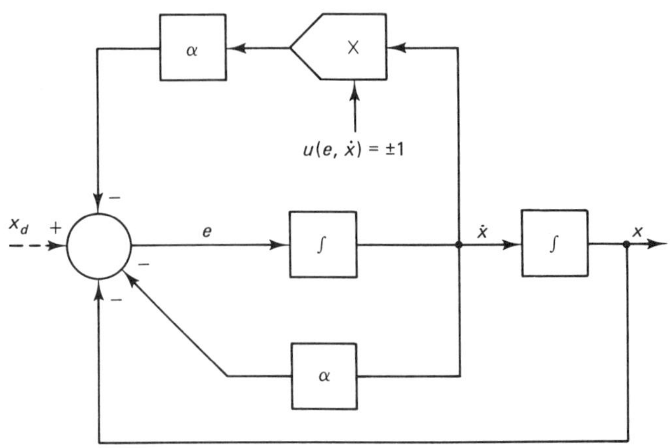

Figure 6.5 Bilinear regulator synthesis, $n = 2$. (From [M2].)

sion (6.68) for λ is dimensionless. Moreover, the switching function p_2x_2 is homogeneous. Therefore, $\mathbf{x}(t)$ and $\mathbf{p}$ can be magnitude scaled by a constant factor ρ_2/ρ_1, and the result is an extremal that terminates on the circle θ_2 with radius ρ_2.

Now consider the synthesis of Fig. 6.3. The extremals corresponding to a circle of radius $k\rho$ are obtained from those shown by performing a uniform contraction (dilation) with scalc factor k on the entire phase plane. In particular, the switching curve Γ of Fig. 6.3 is mapped into a new switching curve $\Gamma_k = \{\mathbf{x} \mid \mathbf{x}/k \in \Gamma\}$.

6.3.3 Suboptimal Control

Suppose that the target circle undergoes an infinite contraction, that is, $k \to 0$. In the limit, θ becomes the origin in the phase plane. As mentioned previously, the optimization problem is not well posed when θ is just the origin since this point is not reachable in finite time. However, the switching curves Γ_k obtained by the contraction process approach a well-defined limit. As $k \to 0$ the switching curve $\Gamma_k = \{\mathbf{x} \mid \mathbf{x}/k \in \Gamma\}$ becomes just the radial line with slope $-\lambda_1$. Thus the limiting extremal synthesis is shown by Fig. 6.6. The plane is divided into sectors R^+ and R^-, with $u(\mathbf{x}) = +1$ on R^+ and $u(\mathbf{x}) = -1$ on R^-. Again, Fig. 6.6 shows the system state diagram.

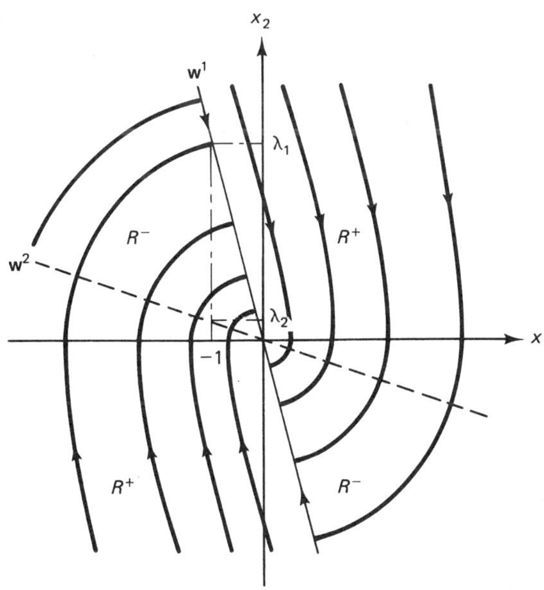

Figure 6.6 Limiting synthesis, $\alpha = 2$. (From [M2].)

The result can be given a very simple interpretation. The eigenvector $\mathbf{w}^2$ corresponding to the smaller eigenvalue λ_2 lies in the dashed line shown in Fig. 6.6. Without the bilinear mode of control (i.e., for the equation $\ddot{x} + 2\alpha x + x = 0$), all trajectories would approach the origin asymptotic to the direction of the "slow" eigenvector $\mathbf{w}^2$. This would result in relatively poor performance. Moreover, as α becomes large the performance deteriorates, since $\lambda_2 \to 0$. With the bilinear mode of control, however, the direction of $\mathbf{w}^2$ lies entirely in the sectors R^-, and the effect of the "slow mode" is nullified. All extremals approach the origin along the "fast" eigenvector $\mathbf{w}^1$. As the damping α becomes large, the eigenvalue λ_1 becomes large and the performance improves monotonically. Unfortunately, it can be shown that the limiting synthesis of Fig. 6.6 is not quite optimal.

It is of interest to compare the performance of the second-order bilinear system with that of a linear system when the input is other than a step. Hence a sinusoidal input with variable frequency is utilized for the bilinear system

$$\ddot{x} + \alpha[1 + u(x - x_d, \dot{x})]\dot{x} + x = x_d, \tag{6.71}$$

where x is the system response for a demanded input x_d. The simple switching law shown in Figs. 6.5 and 6.6 is incorporated for $u(x - x_d, \ddot{x})$. This law was obtained as the limit of syntheses for step-function inputs and was shown to be suboptimal even for this class of inputs. Therefore, no claim is made as regards optimality for the results that are presented.

The control of Fig. 6.6, however, has two nice properties. First, it is easy to synthesize in practice. Also, the nonlinearity is independent of radius in the phase plane, which means that solutions of the resulting system have the important scalar multiplication property. That is, if the forced system (6.71) has the response $x(t)$ for an input $x_d(t)$, it has response $kx(t)$ for the input $kx_d(t)$. Observe that because the switching lines in Fig. 6.6 are radial,

$$u(x - x_d, \dot{x}) = u(kx - kx_d, k\dot{x}),$$

and therefore

$$k\ddot{x} + \alpha[1 + u(kx - kx_d, k\dot{x})]k\dot{x} + kx = kx_d$$

is satisfied.

The solid curves of Fig. 6.7 represent normalized mean-squared tracking error as a function of frequency for the bilinear system with $\alpha = 1$, 2, and 4. Because of the second property discussed previously, these are universal curves that do not depend on the amplitude of $x_d(t)$. The dashed curves represent mean-squared error for the linear system obtained by seting $u = 1$ in (6.71), with $\alpha = 1$, 2, and 4. The performance of the bilinear system is clearly superior, and the trend for increasing α is reversed from that of the linear system.

6.3.4 Time-Optimal Control

The Hamiltonian for the minimum-time control of (6.64) is

$$H(\mathbf{x}, \mathbf{p}, u) = p_0 + p_1 x_2 - p_2[x_1 + \alpha(1 + u)x_2],$$

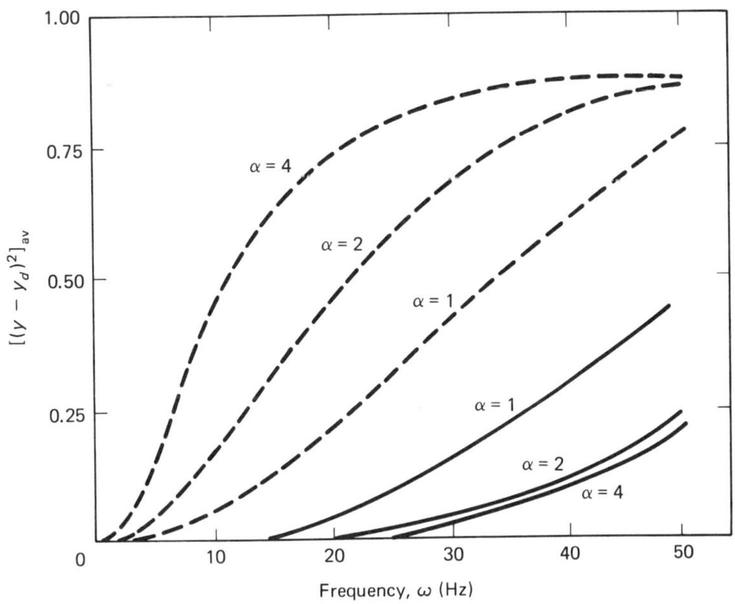

Figure 6.7 Linear versus BLS performance for sinusoidal input. (From [M2].)

and the costate or adjoint system is

$$\begin{aligned} \dot{p}_1 &= p_2 \\ \dot{p}_2 &= -p_1 + \alpha(1 + u)p_2. \end{aligned} \tag{6.72}$$

Along an extremal trajectory, the Hamiltonian is maximized when the control value is

$$u = -\text{sgn}\,(p_2 x_2). \tag{6.73}$$

It is easily verified that no singular arcs exist for this problem.

Now, extremals are generated by computing backward in time from initial points $[x_1(0), x_2(0)]$ on a target circle about the origin in the phase plane. Since the normal to a circle is its radius vector, the transversality condition of the maximum principle is satisfied if we set $p_1(0) = -x_1(0)$, $p_2(0) = -x_2(0)$. Then the value of the Hamiltonian is zero if the constant p_0 is equal to $-2\alpha x_2^2(0)$, which is negative as required by the maximum principle.

The radius of the target circle can, without loss of generality, be taken as unity, since systems (6.64) and (6.72) and relation (6.73) are homogeneous in **x** and **p**. Again, the family of extremals for a target circle of arbitrary radius k can be obtained from those for the unit circle by performing a uniform contraction or dilation with scale factor k in the phase plane.

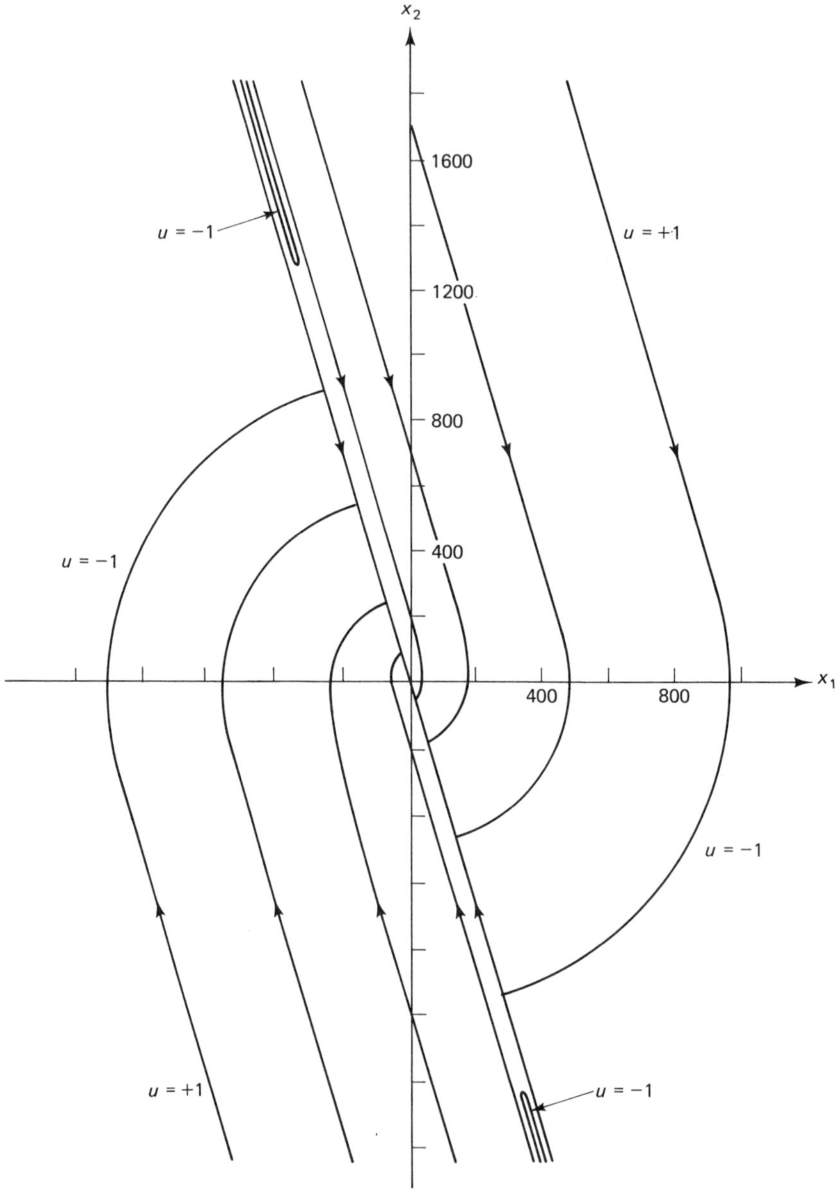

Figure 6.8 Time-optimal phase trajectories, $\alpha = 2$.

Computation of the optimal $\mathbf{x}(t)$ and $\mathbf{p}(t)$ are detailed in [M2]. While no singular solutions exist, there is a small region, as shown by example in Fig. 6.8, where $u = -1$ and $u = +1$ both satisfy the maximum principle. It is a very small region with little loss of performance in using $u = +1$ instead of the $u = -1$, which indeed can be shown to satisfy the cost continuity that was specified for a unique solution in Section 6.3.1 [M2].

The optimal switching law with $\alpha = 2$ is shown in Fig. 6.8, where the unit circle cannot be shown because of the scale. The existence of the large cells with $u = -1$ can be given a simple interpretation. The "slow" eigenvector of the system (6.64) with $u = +1$ lies in these cells, and in order to conserve time the phase should move with zero damping ($u = -1$) to the vicinity of the fast eigenvector before moving toward the origin with maximum damping.

These considerations can be extended to higher-order regulator systems. In classical feedback design, it is well known that rate feedback can stabilize many systems and improve their transient response. On the other hand, an excessive amount of rate feedback causes certain poles (eigenvalues) to approach zero, resulting in a degradation of transient response. Again, this limitation can be circumvented with a bilinear mode of control, since the trajectory can be made to approach its equilibrium state along fast eigenvectors, avoiding those slow ones that correspond to the small eigenvalues.

Example 6.5

Consider time-optimal control of the simplified neutron kinetics

$$\frac{dn}{dt} = \frac{\delta k - \beta}{l} n + \lambda c,$$

$$\frac{dc}{dt} = \frac{\beta}{l} n - \lambda c, \tag{6.74}$$

where n = neutron population
β = portion of neutrons which are delayed (generated from precursors)
λ = average decay constant for an average precursor of population c
l = mean neutron lifetime.

Control of this fission process is affected by means of reactivity of δk. If $u(t) = \delta k(t)$ is constant, the eigenvalues for the system $\mathbf{A}$ matrix,

$$\mathbf{A} = \begin{bmatrix} \dfrac{u - \beta}{l} & \lambda \\ \dfrac{\beta}{l} & -\lambda \end{bmatrix}, \tag{6.75}$$

are

$$\rho_1, \rho_2 = \frac{(u - \beta - \lambda l) \pm [(u - \beta - \lambda l)^2 + 4u\lambda l]^{1/2}}{2l}, \tag{6.76}$$

and the corresponding eigenvectors have directions θ_1 and θ_2 such that

$$\tan \theta_i = \frac{l\rho_i - (u - \beta)}{\lambda l}, \qquad i = 1, 2. \tag{6.77}$$

$x_1 = n$, and $x_2 = c$. The first problem is to find the admissible control $|u(t)| \leq \gamma$ from the class of piecewise continuous functions that drives the system from an initial equilibrium (n_0, c_0) to a terminal equilibrium (n_f, c_f) in minimal time.

Following the maximum principle, the optimal control must maximize the inner product

$$\begin{aligned} \mathcal{H}(n, c, \mathbf{p}, u) &= \mathbf{f}^{\mathrm{T}}(\mathbf{x}, u)\mathbf{p} \\ &= \frac{u}{l}np_1 + (p_2 - p_1)\left(\frac{\beta}{l}n - \lambda c\right), \end{aligned} \tag{6.78}$$

where $\mathbf{p}(t)$ is a continuous nonzero n-vector with

$$\begin{aligned} \frac{d\mathbf{p}}{dt} &= \mathbf{A}^{\mathrm{T}}(u)\mathbf{p}, \\ \mathbf{p}(t_0) &= \mathbf{p}_0. \end{aligned} \tag{6.79}$$

Consequently, since $n(t)$ is positive, the optimal control has the form

$$u = \gamma \operatorname{sgn} p_1(t), \tag{6.80}$$

and the maximum value of (6.78) must be nonnegative, as required by the maximum principle. Substitution of $u = \pm\gamma$ into (6.74), with $(\beta/l)n_0 = \lambda c_0$ for initial equilibrium, yields

$$n = \frac{[\rho_2 \mp (\lambda/l)]n_0 \exp[\rho_1(t - t_0)] - [\rho_1 \mp (\gamma/l)]n_0 \exp[\rho_2(t - t_0)]}{\rho_2 - \rho_1} \tag{6.81}$$

and

$$c = \frac{\rho_2 c_0 \exp[\rho_1(t - t_0)] - \rho_1 c_0 \exp[\rho_2(t - t_0)]}{\rho_2 - \rho_1},$$

where ρ_1 and ρ_2 are given by (6.76) with $u = \pm\gamma$.

Since the two eigenvalues are real and of opposite sign for $u = \gamma$, and both are negative and real for $u = -\gamma$, the bang-bang-controlled neutron kinetics are characterized in the state plane by a saddle-point portrait superimposed on a stable-node portrait with a singular point at the origin. It can be seen by elimination of time from (6.81) that the state trajectories are located by

$$\left[\frac{(\rho_2 + \lambda)c - (\beta/l)n}{(\rho_2 + \lambda)c_0 - (\beta/l)n_0}\right]^{-\rho_2} = \left[\frac{(\rho_1 + \lambda)c - (\beta/l)n}{(\rho_1 + \lambda)c_0 - (\beta/l)n_0}\right]^{-\rho_1}, \tag{6.82}$$

where n_0 and c_0 locate any given state. Certain such combinations of time-optimal trajectories are given in Fig. 6.9. State behavior of the adjoint system (6.79), with $u = 0.9\beta$ for positive p_1 and $u = -0.9\beta$ for negative p_1, is described by Fig. 6.10. This so-called costate in the right half of the state plane is described by a saddle-point portrait, while that in the left half-plane is typical of an unstable node portrait. Since it is required that $\mathbf{p}(t)$ be continuous, it is obvious from Fig. 6.10 or from the solution of (6.79) directly that $p_1(t)$, the switching function in (6.80), cannot have more than one zero in finite time.

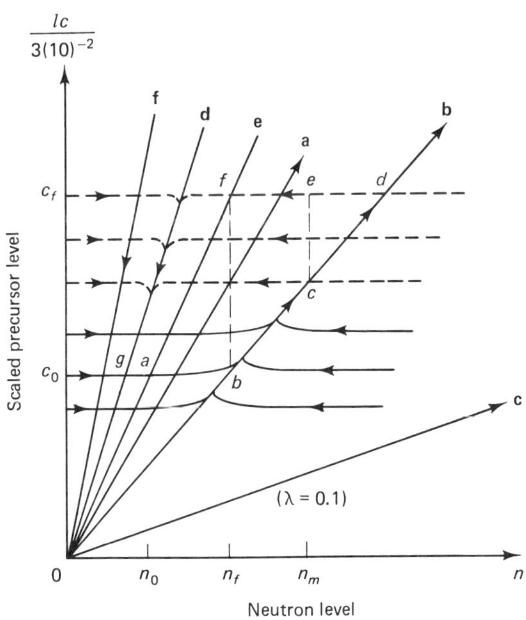

Figure 6.9 Time-optimal neutronic state trajectories. (From [M2].)

Since the constant-u trajectories are unique solutions of the differential equations of the system, Fig. 6.9 indicates that there is only one possible trajectory that joins any reachable state from some initial state with a maximum of one switching, and therefore such a trajectory must be optimal (e.g., trajectory a–d–f has $u = \gamma$ from a to d and $u = -\gamma$ from d to f). Also, the principle of optimality (any portion of an optimal trajectory is optimal) shows that if a desired terminal state can be reached from a given initial state with no switchings and $|u| = \gamma$, the joining trajectory is optimal.

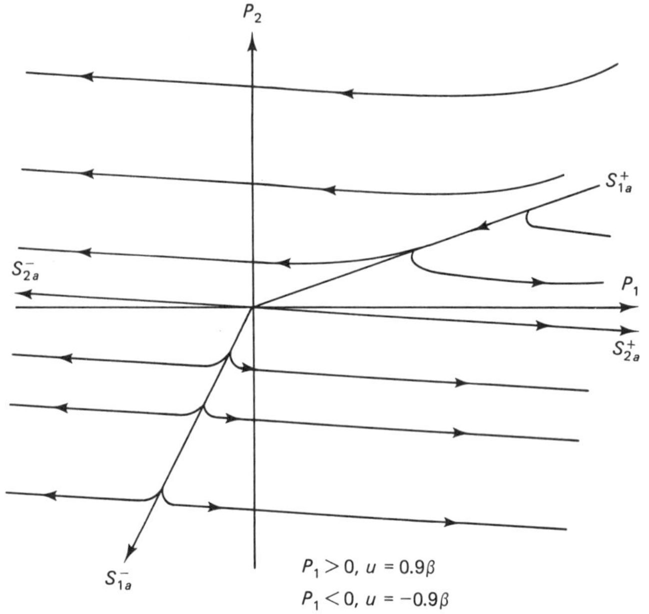

Figure 6.10 Neutronic adjoint trajectories. (From [M2].)

An example of an optimal startup trajectory linking an initial steady state (n_0, c_0) with a desired terminal steady state (n_f, c_f) is represented by trajectory *a–d–f* in Fig. 6.9. Initially, the system is at equilibrium with zero reactivity. Then reactivity is made equal to the maximum positive constant γ while $p_1(t)$ is positive. At point *d*, $p_1(t)$ goes through zero and becomes negative as reactivity is switched to the minimum constraint. When the desired state point *f* is reached, reactivity is returned to zero to maintain quiescence.

The switching point for the time-optimal startup process is conveniently determined from (6.82) by the neutron level n (or the neutron precursor level c) at point *d*. In general, (6.82) may be used to generate a switching line for a time-optimal startup to a desired terminal state (with n_0 replaced by n_f and c_0 by c_f). Below this line, $u = \gamma$ and $u = -\gamma$ above the switching line.

If knowledge of n and c is available, an on-line computation could be made (e.g., digitally, or by means of an analog function generator and a comparator) to determine the switching point.

Since the trajectory is asymptotic to the eigenvector b in Fig. 6.9 for any significant difference $n_f - n_0$, the switching point is nearly independent of the initial state. In many physical systems, it might be desirable to switch to a simple control law that is a continuous function of the system state within some small predetermined region of the desired terminal state. The size and shape of this region could be determined by the accuracy of the equations in defining the physical process, the accuracy of the measuring devices, and the capability of the closed-loop control. Hence the exact switching point might not be too important.

Let us end this chapter with a note that such optimal controls generally are not synthesized in practice, but do provide a design basis. A yardstick for judging performance and criticality of certain design parameters and constraints can be most useful.

EXERCISES

6.1. Compute the time-optimal control $u(\mathbf{x})$ of a simple inertial load with $\dot{x}_1 = x_2$, $\dot{x}_2 = u$ and $|u| \leq 1$ such that $x_1(0) = x_{10}$, $x_2(0) = x_{20}$, and $x_1(t_f) = 0 = x_2(t_f)$.

6.2. Compute the optimal linear regulator, $u(x)$, for $\dot{x} = -\alpha x + bu$ with $J = \int_0^\infty (x^2 + u^2)\,dt$, $x(0) = x_0$, and $x(\infty) = 0$.

6.3. Compute the time-optimal control for $\dot{x} = -x + ux + u$, where $|u| \leq 1$, $x(0) = x_0$, and $x(t_f) = 0$.

6.4. Consider the prompt neutron kinetics (no precursors) such that $\dot{x} = (u/l)x$, where $l = 10^{-4}$ sec, $|u| \leq 0.9$, and $|\dot{u}| \leq n$. Compute the time-optimal control $u(t)$ from $x(0) = x_0$ to $x(t_f) = x_f$.

6.5. Consider the time-optimal control of

$$\dot{x}_1 = x_2,$$

$$\dot{x}_2 = -x_1 - 2(1 + u)x_2,$$

where $|u| \leq 1$, from a general initial state $\mathbf{x}_0$ to a small terminal circle of radius ρ. Show that the solution from the maximum principle verifies Fig. 6.8. Discuss the consequences of nonunique solutions here.

6.6. Repeat Exercise 6.3 with $\dot{x} = -x + ux + v$, where $|u| \leq 1$, $|v| \leq 1$.

6.7. Consider the time-optimal control of the single-precursor nuclear fission process, (6.74) with $\dot{c} = 0$ for $t < 0$, $\dot{n} = 0$ for $t < 0$ and $t > t_f$ and $|u| < \gamma\beta$, where γ is a constant less than 1. Show that the required terminal control $u(n, c)$ satisfies the required inequality. Repeat the problem with consideration of rate constraint $|\dot{u}| \leq \eta$.

6.8. Analyze stability of the desired terminal trajectory with the required terminal control $u(n, c)$ for $t \geq t_f$ and imprecise state measurement in Exercise 6.7. How does this affect the actual control synthesis?

6.9. Consider the scalar nonlinear system $\dot{x} = f(x) + u$, where $x(0) = x_0$, $x(t_f) = x_f$, $|f(x)| \leq \alpha$, and $|u| \leq 1$.

(a) Compute the time-optimal control $u(x)$.

(b) Analyze controllability dependence on α.

CHAPTER 7

DESIGN CASE STUDIES

7.1 INTRODUCTION

Detailed control-system design studies are presented in this chapter as examples of nonlinear, particularly bilinear, applications. The methodologies of Chapter 6 are applied to the control of a surface vehicle and a nuclear rocket engine. Numerous other applications may be found in the references, many of which provide more details of the examples provided throughout this book. Finally, suggested future studies, including adaptive-control applications, are noted.

7.2 SWITCHING-TIME COMPUTATION

It was noted in Section 6.3 that the bang-bang control is at least an extremal process (i.e., it satisfies the maximum principle) for nonlinear systems which are linear in control with a cost function which likewise is linear in control. This may result in a variable structure or adaptive process of high performance. Unfortunately, the two-point, boundary-value problem (TPBVP) that must be solved to find the switching times can be most difficult. Next, an iterative computational method, called the switching-time-variation method (STVM), is developed for a successive approximation of a solution to the TPBVP [M2]. The concept of switching-time variation is introduced to compute the gradient of the cost functional with respect to switching times.

The STVM is a computational technique to obtain optimum bang-bang controls for the problem discussed above by means of a modified gradient scheme. The basic idea of the method is to generate a sequence of switching functions and to compute the gradient of the cost with respect to the swtiching times. With this gradient, the switching times are corrected in each iteration so that the control sequence converges to an optimal nonsingular control.

For convenience, it is assumed that the controls are scalar functions, but the arguments can be generalized easily for vector controls. Then (6.62) and (6.63) become

$$\frac{d\tilde{\mathbf{x}}}{dt} = \tilde{\mathbf{a}}(\tilde{\mathbf{x}}) + \tilde{\mathbf{b}}(\tilde{\mathbf{x}})u, \tag{7.1}$$

where $\tilde{\mathbf{x}}^{\mathrm{T}} = [x_0 \quad \mathbf{x}]$, $x_0(t_f) = J$, $x_0(t_0) = 0$,

$$\tilde{\mathbf{b}}(\tilde{\mathbf{x}}) = \begin{bmatrix} b_0(\mathbf{x}) \\ \mathbf{b}(\mathbf{x}) \end{bmatrix} \quad \text{and} \quad \tilde{\mathbf{a}}(\tilde{\mathbf{x}}) = \begin{bmatrix} a_0(\mathbf{x}) \\ \mathbf{a}(\mathbf{x}) \end{bmatrix}.$$

Also, a free terminal state (i.e., $\theta = \mathbb{R}^n$) and fixed terminal time are assumed to simplify the gradient computations. This does not mean, however, that the method cannot be utilized for problems with a constrained terminal set θ. For these problems, the target set can be considered by means of adjoining a penalty term to the cost functional so that the problem with modified cost converges to the original problem. Free terminal time problems may be solved by means of a sequence of fixed-time problems [Note that (7.1) is a generalization of the BLS.]

7.2.1 Foundations

As a basis for the method, define the switching vectors $\boldsymbol{\tau} \in \mathbb{R}^N$ and $\boldsymbol{\tau}_i \in \mathbb{R}^N$, and the switching set $T \subset \mathbb{R}^N$ so that

$$\boldsymbol{\tau}^N = [\tau_1^N \ . \ . \ . \ \tau_N^N]^{\mathrm{T}}, \tag{7.2}$$

$$\boldsymbol{\tau}_i^N = [\tau_{i1}^N \ . \ . \ . \ \tau_{iN}^N]^{\mathrm{T}}, \tag{7.3}$$

and

$$T^N = \{\boldsymbol{\tau}^N : t_0 \leq \tau_1^N \leq \cdots \leq \tau_N^N \leq t_f\}. \tag{7.4}$$

The elements τ_j^N and $\tau_{ij}^N (j = 1, \ldots, N)$ of $\boldsymbol{\tau}^N$ and $\boldsymbol{\tau}_i^N$ are called switching times. Also, let

$$v(\boldsymbol{\tau}^N) = \sum_{j=0}^{N} (-1)^j [\delta_{-1}(t - \tau_j^N) - \delta_{-1}(t - \tau_{j+1}^N)]$$

and

$$V^N = \{v(\boldsymbol{\tau}^N) : \tau^N \in T^N, N = \text{nonnegative integers}\},$$

where $\delta_{-1}(t)$ is the unit step function. It follows that $V^N \subset U$, and a bang-bang control $u(t) \in V$ such that

$$u(t) = v(\boldsymbol{\tau}^N) \tag{7.5}$$

for some N and some $\boldsymbol{\tau}^N \in T^N$.

Substitution of (7.5) into (7.1) yields

$$\frac{d\tilde{\mathbf{x}}}{dt} = \tilde{\mathbf{a}}(\mathbf{x}) + (-1)^j \tilde{\mathbf{b}}(\mathbf{x}) \tag{7.6}$$

for $t \in [T_j^N, T_{j+1}^N)$, $j = 0, \ldots, N$, such that $\boldsymbol{\tau}_{N+1}^N = t_f$. Obviously, (7.6) has a continuous and unique solution $\tilde{\mathbf{x}}(t)$ on $[T_j^N, T_{j+1}^N)$ from which it can be shown that $\tilde{\mathbf{x}}(t_f)$, and therefore $x_0(t_f) = J$, is continuous in τ^N.

Then define a continuous function I by

$$I(\boldsymbol{\tau}^N) = J(\mathbf{x}_0, v(\boldsymbol{\tau}^N), t_0) = x_0(t_f). \tag{7.7}$$

From the variational form (perturbation equation) of (7.1), with v_i the ith-switching control ($i = 1, \ldots$) in place of u, the gradient of I with respect to the switching vector τ^N may be computed. Consequently,

$$\text{grad } I(\boldsymbol{\tau}_i^N) = \frac{\partial I}{\partial \tau^N}(\boldsymbol{\tau}_i^N) = -\boldsymbol{\phi}_i^N, \tag{7.8}$$

where

$$\boldsymbol{\phi}_i^N = [\phi_{i1}^N \ -\phi_{i2} \ldots (-1)^{N-1}\phi_{iN}^N]^{\mathrm{T}},$$
$$\phi_{ij} = \phi_i(\tau_{ij}^N),$$
$$\phi_i(t) = 2\tilde{\boldsymbol{\lambda}}_i^{\mathrm{T}}(t)\tilde{\mathbf{b}}(\mathbf{x}(t)),$$

and $\tilde{\boldsymbol{\lambda}}_i(t)$ is the solution to the adjoint equation

$$\frac{d\tilde{\boldsymbol{\lambda}}_i}{dt} = -\frac{\partial}{\partial \tilde{\mathbf{x}}}[\tilde{\mathbf{a}}(\mathbf{x}_i) + v_i\tilde{\mathbf{b}}(\mathbf{x}_i)]^{\mathrm{T}}\tilde{\boldsymbol{\lambda}}_i, \tag{7.9}$$

with

$$\tilde{\boldsymbol{\lambda}}_i(t_f) = \left.\frac{\partial I}{\partial \tilde{\mathbf{x}}}\right|_{\mathbf{x}_i(t_f)}.$$

Now the gradient vector may be utilized to generate a nonincreasing sequence of cost functionals and corresponding switching vectors. In this manner let an improved switching vector be given by

$$\boldsymbol{\tau}_{i+1}^N = \boldsymbol{\tau}_i^N + \mathbf{K}_i\boldsymbol{\phi}_i^N, \tag{7.10}$$

where $\mathbf{K}_i$ is an $N \times N$ diagonal matrix with sufficiently small nonnegative elements such that $\boldsymbol{\tau}_{i+N}^N$ belongs to the switching set T^N.

The fact that the gradient procedure above generates a nonincreasing minimizing sequence of cost functionals and corresponding switching vectors may be shown from compactness (closed and boundedness) of the sets T^N and $I(T^N)$. Then, by substitution, it can be shown that the STVM solution above satisfies the TPBVP for the maximum principle. That is, the solution to (7.9) approaches the costate solution of the maximum principle, $\tilde{\boldsymbol{\lambda}}(t) \to \tilde{\mathbf{p}}(t)$, and $\tilde{\mathbf{x}}(t)$, the solution to (7.6), approaches the solution to (7.1) such that the Hamiltonian becomes a maximum. See [M2] for more details.

In summary, the maximum principle for (7.1) is given as follows: There exists a continuous nonzero costate $\tilde{\mathbf{p}}(t)$, an extremal control $u(t)$, such that

$$\begin{aligned} H(\mathbf{x}, \mathbf{p}, u) &= \mathbf{p}^{\mathrm{T}}[\mathbf{a}(\mathbf{x}) + \mathbf{b}(\mathbf{x})u] - [a_0(\mathbf{x}) + b_0(\mathbf{x})u] \\ &= \tilde{\mathbf{p}}^{\mathrm{T}}[\tilde{\mathbf{a}}(\mathbf{x}) + \tilde{\mathbf{b}}(\mathbf{x})u] \end{aligned} \tag{7.11}$$

is maximized by

$$u = \operatorname{sgn} s(t), \tag{7.12}$$

where $p_0 = -1$ in (6.5),

$$s(t) = \tilde{\mathbf{b}}^{\mathrm{T}}(\mathbf{x})\tilde{\mathbf{p}}, \tag{7.13}$$

and

$$\frac{d\mathbf{p}}{dt} = -\frac{\partial H}{\partial \mathbf{x}}. \tag{7.14}$$

If $\mathbf{x}(t_f)$ is constrained to the terminal set θ, then $\mathbf{p}(t_f)$ is orthogonal to θ.

If $s(t)$ is nonzero for all finite time intervals on $[0, t_f]$, the STVM may be used to compute the optimal switching times. Also, the method may be used to approximate singular solutions.

Although the STVM is presented for scalar control, the procedure is applicable to vector control as well. Also, the method, as presented above, assumes knowledge of the optimal number of switchings. A similar gradient procedure, however, may be utilized to compute the optimal number of switchings [M2]. Also, the foregoing summary assumes a variable terminal time. But a simple modification with $T_{N+1}^N = t_f$ and $\delta t_f = 0$ accommodates the fixed terminal-time case.

In summary, the algorithm consists of the following steps:

1. Guess a number of switchings N and a switching vector $\boldsymbol{\tau}_i^N$.
2. Obtain the modified state equation (7.6) by substituting $v(\boldsymbol{\tau}_i^N)$ into the original state equation (7.1). Then, starting from $\tilde{\mathbf{x}}(t_0) = \tilde{\mathbf{x}}_0$, integrate (7.6) from t_0 to t_f; t_f can be obtained sequentially from the desired terminal condition, or guessed, for the first iteration if no stopping condition is given. Although the initial choice of the terminal time t_f is not usually critical, it is desirable to choose the value as close as possible to optimal to reduce computation time. Store the resulting state trajectory $\tilde{\mathbf{x}}_i(t)$, for $t \in [t_0, t_f]$, and the switching vector $\boldsymbol{\tau}_i^N$.
3. Compute the adjoint trajectory $\tilde{\boldsymbol{\lambda}}_i(t)$ by integrating (7.9) backward in time using $\mathbf{x}_i(t)$ from step 2. Store the values of the adjoint vector $\tilde{\boldsymbol{\lambda}}_i(t)$ for $t = \tau_{ij}^N (j = 1, \ldots, N)$.
4. Compute the gradient vector $-\boldsymbol{\phi}_i^N$ and compute $\delta\tau_{ij}^N = k_{ij}\phi_{ij}^N$ by choosing a reasonable $k_{ij} (j = 1, \ldots, N)$. Note that if the k_{ij} are picked too small, the convergence of the solution is very slow; on the other hand, if the k_{ij} are picked too large, the condition $\boldsymbol{\tau}_i^N + \mathbf{K}_i\boldsymbol{\phi}_i^N \in T^N$ may not be satisfied.
5. Compute a new switching vector $\boldsymbol{\tau}_{i+1}^N$ from $\boldsymbol{\tau}_i^N + \delta\boldsymbol{\tau}_i^N$.
6. Repeat steps 2 to 5 until $\Sigma_{j=1}^N (\delta\tau_{ij}^N)^2 < \epsilon$, where $\epsilon > 0$ is a small given number.
7. Check if any of the intervals between the switching times approaches zero. If so, discard those switching times and renumber the switching times and the number of switchings.
8. Compute $\boldsymbol{\phi}(t)$ and check if any zeros of $\boldsymbol{\phi}^T = [\phi_0(t) \quad \ldots \quad \phi_N(t)]$ are between two switching times. If so, increase the number of switchings and accordingly redefine the switching vector.
9. Repeat steps 2 to 8 until all the zeros of $\boldsymbol{\phi}(t)$ coincide with the switching times.
10. Increase the number of switchings, and repeat steps 2 to 9 until the cost functional cannot be decreased by increasing the number of switchings.
11. Print the final switching times, the state trajectory, and the cost.

Example 7.1 [M2]

Consider a torque-controlled inertial load with independently controlled rate feedback such that

$$\dot{x}_1 = x_2,$$
$$\dot{x}_2 = -(1 + u_2)x_2 + 2u_1,$$

where x_1 is angular position, $\mathbf{x}(t_0) = \mathbf{x}_0$, and $|u_i| \leq 1 (i = 1, 2)$. The problem is to control the system to the origin in minimal time with an admissible control. It follows from the maximum principle that extremal controls are

$$u_1 = \text{sgn}\,(p_2(t)) \qquad \text{and} \qquad u_2 = \text{sgn}\,(p_2(t)x_2(t)),$$

where $p_2(t)$ is an adjoint solution.

It can be shown that the problem does not have a singular solution, that $p_2(t)$ has at most one zero, and that $p_2(t)$ has a maximum of two zeros. Then Fig. 7.1 shows the switching times and cost as a function of iteration for the solution of this problem by the STVM. It is found approximately that $\mathbf{u}^T = [-1, 1]$ on (0, 1.2 sec), $\mathbf{u}^T = [-1, 1]$ on (1.2, 3.8 sec), and $\mathbf{u}^T = [+1, +1]$ on (3.8, 4.8 sec). Here $x_1(0) = 5$ and $x_2(0) = 10$.

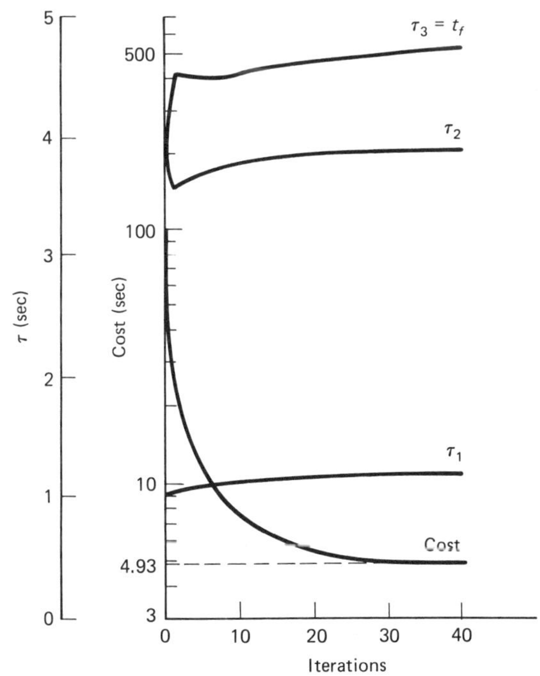

Figure 7.1 STVM cost and switching times versus iteration for minimum-time control.

7.2.2 Fuel-Optimal Control

In this section consider the "fuel-optimal" (weighted with time-optimal) control of the nonlinear system with linear control (6.62). It is desired to drive the state from $\mathbf{x}(t_0) = \mathbf{x}_0$ to $\mathbf{x}(t_f) \in \theta$ with an admissible control $|u(t)| \leq 1$ so that

$$J = \int_{t_0}^{t_f} (e_0 + e_1 |u|)\, dt \tag{7.15}$$

is minimized, where e_0 and e_1 are positive weighting constants.

Again, the form of optimal control maximizes the Hamiltonian

$$H(\mathbf{x}, \mathbf{p}, u) = -(e_0 + e_1 |u|) + \mathbf{p}^{\mathrm{T}}[\mathbf{a}(\mathbf{x}) + \mathbf{b}(\mathbf{x})u]. \tag{7.16}$$

Hence the "fuel-optimal" control

$$u = \operatorname{dez}(\alpha), \tag{7.17}$$

where

$$\operatorname{dez}(\alpha) = \begin{cases} \operatorname{sgn}(\alpha) & \text{if } |\alpha| > 1 \\ 0 & \text{if } |\alpha| < 1, \end{cases} \tag{7.18}$$

$$\alpha = \sum_{j=1}^{n} \frac{b_j p_j}{e_1},$$

and b_j is the jth element of $\mathbf{b}(\mathbf{x})$. It is assumed that $\alpha(t)$ does not vanish on any finite-time interval. Since

$$\operatorname{dez}(\alpha) = \tfrac{1}{2}[\operatorname{sgn}(\alpha - 1) + 1] + \tfrac{1}{2}[\operatorname{sgn}(\alpha + 1) - 1],$$

it follows that (7.17) can be expressed by

$$u(t) = \tfrac{1}{2}\Bigg\{\sum_{j=1}^{N^+} [\delta_{-1}(t - \tau_{2j-1}^+) - \delta_{-1}(t - \tau_{2j}^+)] - \sum_{j=1}^{N^-} [\delta_{-1}(t - \tau_{2j-1}^-) - \delta_{-1}(t - \tau_{2j}^-)]\Bigg\}, \tag{7.19}$$

where τ_j^+, $j = 0, \ldots, 2N^+$, are the zeros of $\alpha(t) - 1$, and τ_j^-, $j = 0, \ldots, 2N^-$, are the zeros of $\alpha(t) + 1$. The control may be specified by the switching vectors of $(2N^+) + (2N^-)$ dimension,

$$\boldsymbol{\tau}^N = \begin{bmatrix} \boldsymbol{\tau}^{N^+} \\ \boldsymbol{\tau}^{N^-} \end{bmatrix}, \tag{7.20}$$

where

$$\boldsymbol{\tau}^{N^+} = \begin{bmatrix} \tau_1^{N^+} \\ \vdots \\ \tau_{2N^+}^{N^+} \end{bmatrix}, \qquad \boldsymbol{\tau}^{N^-} = \begin{bmatrix} \tau_1^{N^-} \\ \vdots \\ \tau_{2N^-}^{N^-} \end{bmatrix}.$$

The switching set T^N is defined by

$$T^N = \{\boldsymbol{\tau}^N : \boldsymbol{\tau}^{N^+} \in T^{N^+} \text{ and } \boldsymbol{\tau}^{N^-} \in T^{N^-}\}, \tag{7.21}$$

where

$$T^{N^+} = \{\boldsymbol{\tau}^{N^+} : t_0 \le \tau_i^{N^+} \le \cdots \le \tau_{2N^+}^{N^+} \le t_f\},$$
$$T^{N^-} = \{\boldsymbol{\tau}^{N^-} : t_0 \le \tau_i^{N^-} \le \cdots \le \tau_{2N^-}^{N^-} \le t_f\}.$$

A control

$$v = v(\boldsymbol{\tau}^N) = v(\boldsymbol{\tau}^{N^+}) + v(\boldsymbol{\tau}^{N^-}), \tag{7.22}$$

where

$$v^+ = v(\boldsymbol{\tau}^{N^+}) = \sum_{j=1}^{N^+} [\delta_{-1}(t - \tau_{2j-1}^{N^+}) - \delta_{-1}(t - \tau_{2j}^{N^+})],$$
$$v^- = v(\boldsymbol{\tau}^{N^-}) = \sum_{j=1}^{N^-} [\delta_{-1}(t - \tau_{2j-1}^{N^-}) - \delta_{-1}(t - \tau_{2j}^{N^-})],$$

and a control set

$$V^N = \{v(\boldsymbol{\tau}^N) : \boldsymbol{\tau}^N \in T^N,\ N = 1, 2, \ldots\}$$

are utilized for $u(t)$ and U in (7.1). Then the system equations are

$$\begin{aligned} \frac{d\tilde{\mathbf{x}}}{dt} &= \tilde{\mathbf{a}}(\mathbf{x}) && \text{for } t \in \{t : v = 0\} \\ \frac{d\tilde{\mathbf{x}}}{dt} &= \tilde{\mathbf{a}}(\mathbf{x}) + \tilde{\mathbf{b}}(\mathbf{x}) + \mathbf{e} && \text{for } t \in \{t : v = +1\} \\ \frac{d\tilde{\mathbf{x}}}{dt} &= \tilde{\mathbf{a}}(\mathbf{x}) - \tilde{\mathbf{b}}(\mathbf{x}) + \mathbf{e} && \text{for } t \in \{t : v = -1\}, \end{aligned} \tag{7.23}$$

where $\tilde{\mathbf{x}}$, $\tilde{\mathbf{a}}(\mathbf{x})$, and $\tilde{\mathbf{b}}(\mathbf{x})$ are defined by (7.1) and (7.17) with

$$\mathbf{e} = \begin{bmatrix} e_1 \\ 0 \\ \vdots \\ 0 \end{bmatrix}, \qquad \mathbf{e} \in \mathbb{R}^{n+1}.$$

Now the switching-time-variation method (STVM), which was introduced above, may be used to compute the optimal control policy in the absence of singular solutions. Again, it can be shown that the resulting minimizing sequence is convergent to an extremal solution of the maximum principle [M2].

Here the variation in cost is computed by

$$\delta J = \left(\frac{\partial J}{\partial \tilde{\mathbf{x}}}\right) \delta \mathbf{x} \bigg|_{t_f} = \int_{t_0}^{t_f} \boldsymbol{\lambda}^{\mathrm{T}}[\tilde{\mathbf{b}}(\mathbf{x})\, \delta v + \mathbf{e}\delta|v|]\, dt,$$

where $\delta v = \delta v^+ + \delta v^-$, $\delta|v| = \delta v^+ - \delta v^-$,

$$\delta v^+ = -\sum_{j=1}^{N^+} [\Delta(t - \tau_{2j-1}^+)\, \delta\tau_{2j-1}^+ - \Delta(t - \tau_{2j}^+)\, \delta\tau_{2j}^+],$$

$$\delta v^- = \sum_{j=1}^{N^-} [\Delta(t - \tau_{2j-1}^-)\, \delta\tau_{2j-1}^- - \Delta(t - \tau_{2j}^-)\, \delta\tau_{2j}^-],$$

and N superscripts are dropped for convenience. Here $\Delta(\cdot)$ is the Dirac delta function [instead of $\delta(\cdot)$] to avoid confusion with δ variation. Substitution and condensation of terms yield

$$\delta J = -\boldsymbol{\phi}^{\mathrm{T}} \delta \boldsymbol{\tau}, \tag{7.24}$$

where $\boldsymbol{\phi} = [\phi_1 \ldots \phi_N]^{\mathrm{T}}$ and

$$\phi_j = (-1)^{j-1} e_1(\beta_j^+ + 1), \qquad j = 1, \ldots, 2N^+,$$

$$\phi_{j+2N} = (-1)^j e_1(\beta_j^- - 1), \qquad j = 1, \ldots, 2N^-,$$

$$\beta_j^+ = \frac{\tilde{\boldsymbol{\lambda}}^{\mathrm{T}}(t)\tilde{\mathbf{b}}(\mathbf{x}(t))}{e_1}\bigg|_{t=\tau_j^+}, \qquad j = 1, \ldots, 2N^+,$$

$$\beta_j^- = \frac{\tilde{\boldsymbol{\lambda}}^{\mathrm{T}}(t)\tilde{\mathbf{b}}(\mathbf{x}(t))}{e_1}\bigg|_{t=\tau_j^-}, \qquad j = 1, \ldots, 2N^-.$$

The gradient of I with respect to the switching vector is given by

$$\text{grad}\, I(\boldsymbol{\tau}^N) = -\boldsymbol{\phi}, \tag{7.25}$$

and the nonincreasing sequence of cost functionals and corresponding switching vectors can be computed from

$$\boldsymbol{\tau}_{i+1} = \boldsymbol{\tau}_i + \mathbf{K}_i \boldsymbol{\phi}_i, \tag{7.26}$$

where $\boldsymbol{\tau}_i$ is the current switching vector, $\boldsymbol{\tau}_{i+1}$ the improved switching vector, $\boldsymbol{\phi}_i$ the negative of the gradient vector, and $\mathbf{K}_i$ an $(N \times N)$-dimensional diagonal matrix with sufficiently small nonnegative diagonal elements such that $\boldsymbol{\tau}_i + \mathbf{K}_i \boldsymbol{\phi}_i \in T^N$.

7.3 CASE STUDIES

Applications to automobile control and nuclear reactor control are studied here as examples of typical nonlinear systems.

7.3.1 Surface Vehicle Control

It is desired to compute the optimal control policy for an automobile (or other surface vehicle) under the influence of conventional frictional braking and acceleration control. First, let us derive a dynamical model for such a system.

Figure 7.2 shows the control force u_1 acting on the brake drum. Since the frictional force is nearly proportional to the orthogonal force between the rubbing surfaces and their relative velocity, the braking force commonly is approximated by

$$f_b = c_b u_1 \dot{x},$$

where $\dot{x}$ is the vehicle translational speed, u_1 the brake pedal control force, and c_b a positive constant. Similarly, other frictional forces may be approximated by

$$f_c = c_f \dot{x},$$

where c_f is a positive constant.

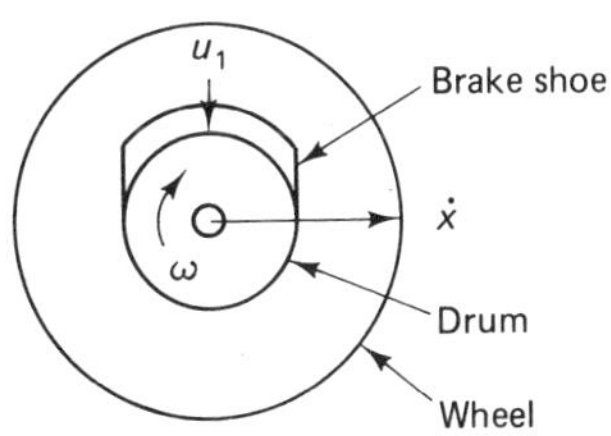

Figure 7.2 Bilinear mechanical brake.

Then a summation of inertial force, frictional forces, and normalized engine force u_2 yields the equations of motion

$$\frac{dx_1}{dt} = x_2,$$
$$\frac{dx_2}{dt} = \frac{-kc_f}{m}x_2 - \frac{kc_b}{m}u_1x_2 + u_2, \tag{7.27}$$

where m is vehicle mass and k is a positive constant. It is seen that the system is bilinear with a parameteric braking force and an additive accelerator force. This checks (2.1).

It is assumed for convenience that the admissible set U is defined by $|u_1| \leq 1$ and $|u_2| \leq 1$; let $a = -kc_f/m = -kc_b/m$. It is a simple matter, however, to let $u_1' = 2u_1 - 1$ with $|u'| \leq 1$, so that physical constraints of the form $0 \leq u \leq 1$ are symmetrical.

The problem considered here is to drive the system (7.27) from an initial state $\mathbf{x}(t_0) = \mathbf{x}_0$ to the origin in the state space so that a weighted fuel and time performance,

$$J = \int_{t_0}^{t_f} (1 + e_1|u_2|)\, dt, \qquad \text{where } e_1 > 0, \tag{7.28}$$

is a minimum. The Hamiltonian and the costate equation for this problem are

$$H = -(1 + e_1|u_2|) + p_1x_2 + p_2[ax_2(1 + u_1) + u_2] \tag{7.29}$$

and

$$\dot{p}_1 = 0,$$
$$\dot{p}_2 = -p_1 - a(1 + u_1)p_2. \tag{7.30}$$

From (7.27) and the maximum principle, with $p_0 = -1$,

$$u_1 = \operatorname{sgn}(s_1) \qquad \text{if } s_1 \neq 0 \tag{7.31}$$

and

$$u_2 = \operatorname{dez}(s_2) \qquad \text{if } |s_2| \neq 1, \tag{7.32}$$

where u_1 and u_2 are extremal controls. The "switching" functions are

$$s_1 = ap_2(t)x_2(t) \tag{7.33}$$

and

$$s_2 = \frac{p_2(t)}{e_1}. \tag{7.34}$$

It is shown now that this "fuel-optimal" problem docs not have a singular solution, except along the trivial trajectory given by $x_1 = x_2 = 0$.

For u_1 to be singular, s_1 must vanish on some nonzero interval. That is, either $p_2(t)$ or $x_2(t)$ must be zero on a nonzero interval.

Suppose that $x_2 = 0$. Then it follows that $\dot{x}_1 = x_2 = 0$, which implies that the system is in equilibrium state. Therefore, assume that $x_2 \neq 0$. Then $p_2(t)$ must be zero for a singular solution. If $p_2 = 0$, then $\dot{p}_2$ must be zero also, and hence

$$\dot{p}_2 = -p_1 - a(1 + u_1)p_2 = -p_1 = 0.$$

But this is impossible according to the maximum principle. Therefore, optimal u_1 must be bang-bang, except along the trivial trajectory $x_1 = x_2 = 0$.

Now, in order for u_2 to be singular, $|s_2(t)| = 1$ must be satisfied on some nonzero interval. That is, either $s_2 = -1$ or $s_2 = +1$. Suppose that $s_2 = -1$. Then, from (7.34),

$$p_2 = -e_1 = \text{constant}. \tag{7.35}$$

Hence $\dot{p}_2 = -p_1 - a(1 + u_1)p_2 = 0$, and

$$p_1 = -a(1 + u_1)p_2 = ae_1(1 + u_1). \tag{7.36}$$

Substitution of (7.35) and (7.36) into (7.29) yields

$$H = -(1 + e_1|u_2| + e_1u_2). \tag{7.37}$$

Then from $s_2(t) = -1$ and from (7.37), it can be seen that $-1 \leq u_2 \leq 0$, and that

$$H = -1 + e_1u_2 - e_1u_2 = -1 < 0.$$

But according to the maximum principle, the extremal Hamiltonian H for a free-terminal-time problem must be zero. Hence the switching function, $s_2(t) = -1$, cannot give an extremal singular solution.

For $s_2(t) = +1$ it is seen that

$$H = -1 - e_1(|u_2| - u_2).$$

But since $s_2(t) = +1$, it follows that $0 \leq u_2 \leq 1$ and that $H = -1 < 0$. Hence the switching function $s_2(t) = +1$ cannot be an extremal switching function either. Hence a nontrivial singular solution does not exist.

It is readily seen that $x_2(t)$ and $p_2(t)$ each can have at most one zero on (t_0, t_f) and that $p_2(t)$ is monotonic. Therefore, $s_1 = ax_2p_2$ can have at most two zeros on (t_0, t_f), and $s_2 = p_2(t)/e_1$ is monotonic and can have at most two switchings. Also, it can be seen from $s_1 = ax_2p_2$ and $s_2 = p_2/e_1$ that the zero of s_1, which corresponds to $p_2 = 0$, must lie between the two switchings of s_2. That is, if $p_2(t_c) = 0$, $|s_1(t_b)| = |s_1(t_d)| = 1$, and $t_b < t_d$, the condition

$$t_b < t_c < t_d$$

must hold. Hence the maximum number of optimum switchings for the fuel-optimal control example is four.

At this point it can be assumed that the forms of extremal controls and state trajectories are given since the solution of (7.27) is well known for such piecewise constant controls. Now the exact locations of the switching times are computed by using the switching-time-variation method. For numerical computations, assume that $t_0 = 0$, $a = 1$, $k/m = 10$, $x_{10} = 5$, $x_{20} = 10$, and $e_1 = 1$.

For the numerical values given above, it can be shown that the optimum number of switchings is four, and the extremal control is given by

$$\begin{aligned} u_1(t) &= \delta_{-1}(t - t_0) - 2\delta_{-1}(t - \tau_1) + 2\delta_{-1}(t - \tau_3) - \delta_{-1}(t - t_f), \\ u_2(t) &= -\delta_{-1}(t - t_0) + \delta_{-1}(t - \tau_2) + \delta_{-1}(t - \tau_4) - \delta_{-1}(t - t_f), \end{aligned} \tag{7.38}$$

where $\delta_{-1}(t)$ is the unit step function, $\tau_1 = t_a$, $\tau_2 = t_b$, $\tau_3 = t_c$, and $\tau_4 = t_d$.

To start the numerical computations, it is initially assumed that $\boldsymbol{\tau}_0 = [1.0\ 2.0\ 3.0\ 4.0\ 5.0]^T$, where $\tau_5 = t_f = 5.0$.

Application of the STVM, with the initial value and the control given above, shows that the sequence of the switching vector $\{\boldsymbol{\tau}_i\}$ converges to $[1.2\ 3.64\ 3.78\ 4.54\ 4.83]^T$. The final cost corresponding to this switching vector is 8.91. The numerically computed results are shown in Table 7.1 and Figs. 7.3 to 7.5. With no consideration to minimal-time performance, $u_2 = 0$ for $t \in (t_0, t_f)$.

A simple transformation of control variables accommodates the nonsymmetrical constraint case, and the problem does exemplify the application of the STVM to fuel-optimal control. In practice, additive braking (i.e., negative of accelerating force) may be synthesized by means of a controlled reversed thrust similar to that used by high-performance landing aircraft. On the other hand, the positive multiplicative control u_2 might evolve from a feedback multiplicative synthesis of an additive control.

Passenger car and commercial vehicle braking performances are studied in [O2], [P5], and [T2]. For safety purposes, maximum deceleration is limited to approximately 15 to 18 ft/sec^2, depending on the vehicle weight, and pedal force is limited to 200 lb, which corresponds to $u_1 \leq \gamma_1$.

TABLE 7.1 NUMERICAL RESULTS FOR "FUEL-OPTIMAL" CONTROL

ITERATION NUMBER 0						
TAU	0.0	0.1000E 01	0.2000E 01	0.3000E 01	0.4000E 01	0.5000E 01
X1(T)	0.5000E 01	0.8756E 01	0.8244E 01	0.6733E 01	0.6080E 01	0.6559E 01
X2(T)	0.1000E 02	0.4887E 00	−0.1511E 01	−0.1511E 01	0.2045E 00	0.8370E 00
COST =	0.2266E 03					
PHI(T)	0.0	0.7991E 02	0.3907E 03	0.4889E 02	0.1190E 03	−0.5963E 02
ITERATION NUMBER 10						
X1(T)	0.5000E 0.	0.8801E 01	0.4210E 01	0.3534E 01	0.1664E 01	0.1609E 01
X2(T)	0.1000E 02	0.4364E −01	−0.4285E 01	−0.4285E 01	−0.5453E 00	0.1121E −01
COST =	0.2126E 02					
PHI(T)	0.0	0.1945E 01	0.3875E 02	0.3170E 02	0.1287E 02	−0.2402E 01
ITERATION NUMBER 20						
TAU	0.0	0.1187E 01	0.3556E 01	0.3698E 01	0.4571E 01	0.4776E 01
X1(T)	0.5000E 01	0.8801E 01	0.3197E 01	0.2573E 01	0.6190E 00	0.5169E 0
X2(T)	0.1000E 02	0.2355E −01	−0.4735E 01	−0.4735E 01	1.8262E 00	−0.2166E 00
COST =	0.1011E 02					
PHI(T)	0.0	0.3536E 00	0.9881E 01	0.9659E 01	0.1137E 01	0.4222E 01

Source: Ref. [M2].

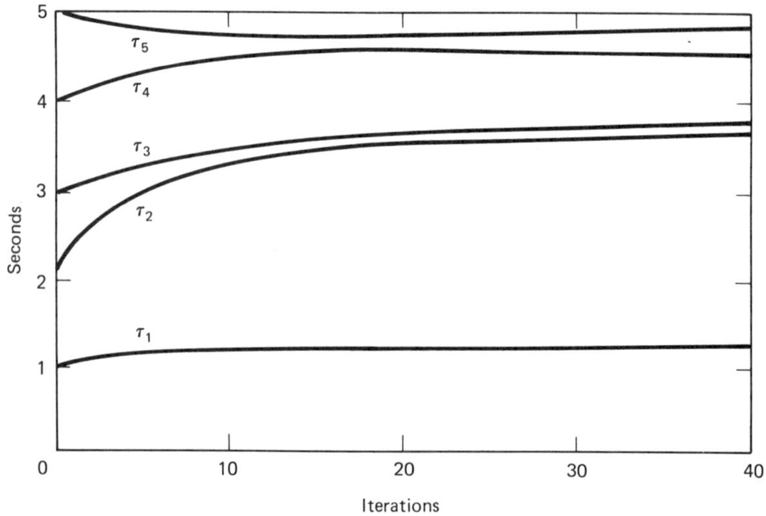

Figure 7.3 Switching times versus iteration: fuel-optimal control.

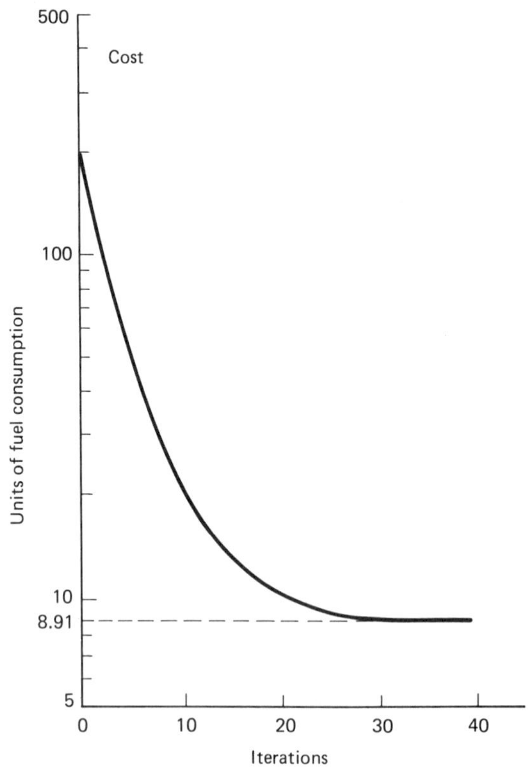

Figure 7.4 Fuel consumption versus iteration for Fig. 7.3.

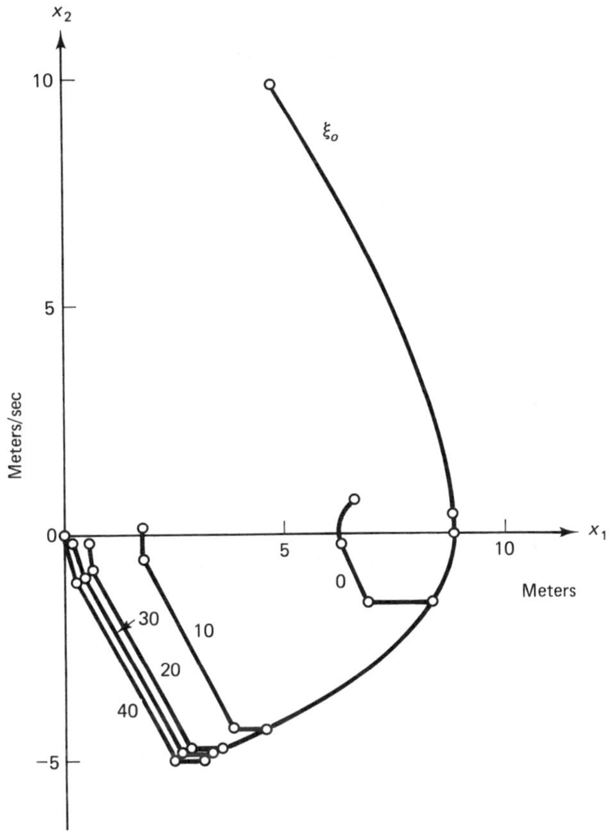

Figure 7.5 Sequential fuel-optimal trajectories for STVM. Numbers along trajectories refer to number of iterations.

If maximum deceleration is limited to $\dot{x}_2 = -d$ in (7.27), the corresponding controls can be computed from

$$d = (a + nu_1)x_2 - u_2,$$

where $a = -kc_f/m$ and $n = -kc_b/m$; u_1 is the friction brake control, and u_2 is the normalized acceleration control. If u_1 and u_2 are nonnegative, and furthermore $u_2 = 0$ during braking, the required braking control $u_1 \to \infty$ as $x_2 \to 0$. But $u_1 \leq \gamma_1$. Consequently, the braking policy, even with deceleration constraint, can be improved by a normalized additive braking force $u_3 \geq 0$. Then the maximum allowable deceleration (to minimize stopping distance) is realized in $t \in (t_0, t_f)$ by

$$u_3 = d - (a + n\gamma_1)x_2. \tag{7.39}$$

If a large frontal area is projected to add an appreciable drag force, the braking force is again a function of speed. Since the drag force is approximately proportional to the dynamic pressure multiplied by the frontal area A_d,

$$f_d = c_d A_d x_2^2.$$

If the frontal area is considered a control variable (as with aircraft wing flaps), the system apparently becomes nonlinear in state but linear in control.

With friction braking, reversed-thrust braking, and controlled drag, (7.27) becomes

$$\frac{dx_1}{dt} = x_2,$$

$$\frac{dx_2}{dt} = -(a + n\gamma_1 + bx_2u_4)x_2 + u_2 - u_3, \tag{7.40}$$

where $b = c_d/m$, and u_4 represents frontal drag area. For emergency stops, u_4 might be synthesized by drag chutes.

With such added drag, (7.39) becomes

$$u_3 = d - (a + n\gamma_1 + bx_2u_4)x_2.$$

Figures 7.6 and 7.7 show typical control and speed braking trajectories versus vehicle position for additive braking control u_3 and for drag control u_4, respectively. Note that friction braking u_1 reaches a magnitude constraint in the latter case, and the allowable deceleration is not maintained. Consequently, the reversed-thrust control u_3 is more effective.

Figure 7.8 shows the effect of a terminal pulse of constrained magnitude γ_3 and of normalized impulse $\gamma_3 T$, where T is the pulse length. With impulse equal to change in momentum

$$\gamma_3 T = x_2(t_f - T)$$

is required to stop the vehicle at designated $t = t_f$, which might correspond to some predetermined terminal position.

Another useful variable-structure or adaptive control application to surface vehicles uses active or passive control of suspension-system damping. Figure 7.9 presents a three-lump mechanical-network model for this system. Motion is taken relative to some surface reference with the masses m_1, m_2, m_3 in equilibrium. y_1, y_2, y_3 are positions of the vehicle sprung mass m_1, of the passenger mass m_2, and of the vehicle unsprung mass m_3. Then a force balance on each mass formulates the following

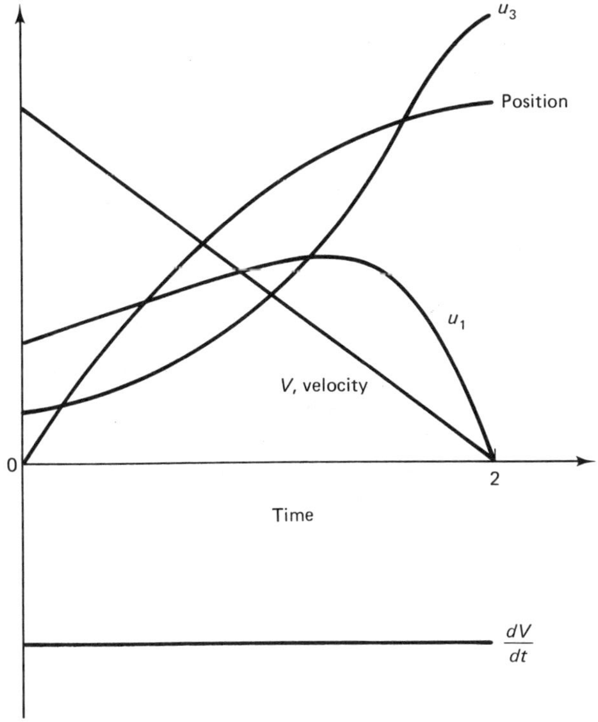

Figure 7.6 Additive and multiplicative braking force to minimize stopping distance.

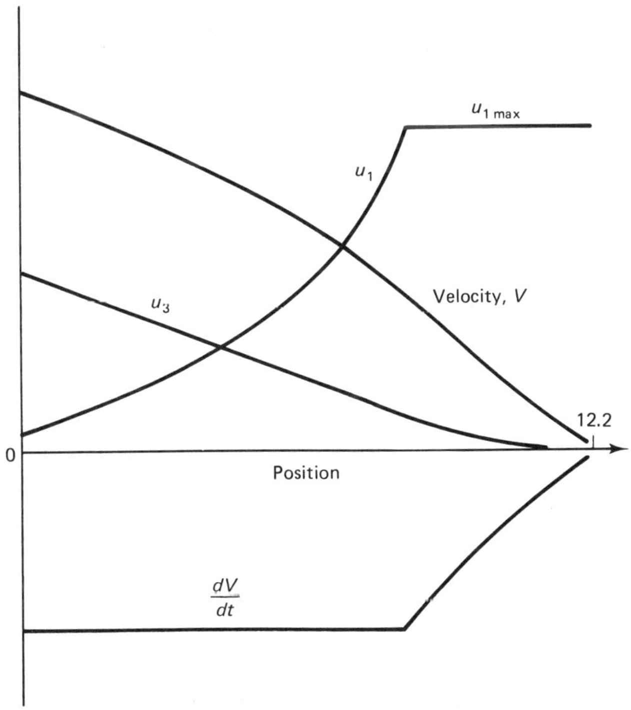

Figure 7.7 Multiplicative and drag braking trajectories.

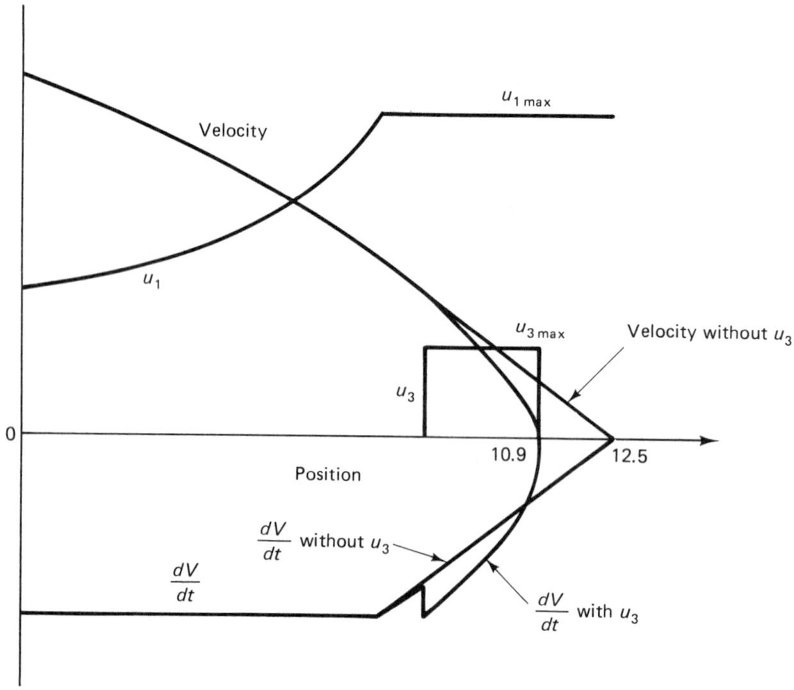

Figure 7.8 Multiplicative and multiplicative plus terminal pulse additive brake comparison.

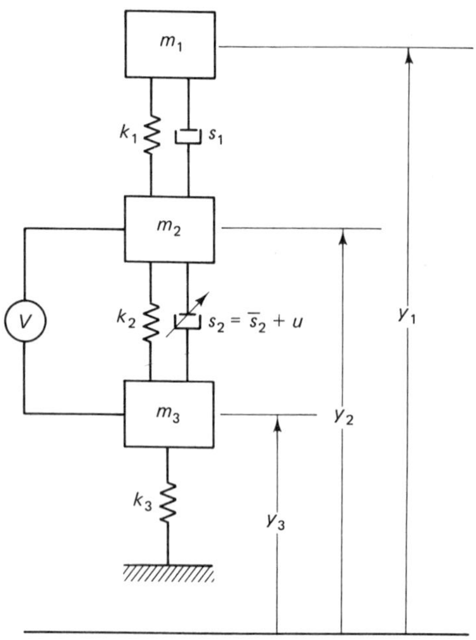

Figure 7.9 Vehicle suspension system.

model:

$$
\begin{aligned}
m_1\ddot{y}_1 &= -k_1(y_1 - y_2) - s_1(\dot{y}_1 - \dot{y}_2), \\
m_2\ddot{y}_2 &= -k_1(y_2 - y_1) - k_2(y_2 - y_3) - s_1(\dot{y}_2 - \dot{y}_1) \\
&\quad - s_2(y_2 - y_3) + v, \\
m_3\ddot{y}_3 &= -k_2(y_3 - y_2) - k_3(y_3 - r) - s_2(\dot{y}_3 - \dot{y}_2) - v,
\end{aligned}
\tag{7.41}
$$

where v is a position control variable and r is a surface input displacement term which directly affects x_3 (vertical position of the wheels, etc.). With controlled active displacement v, the process is a linear control system. If s_2 is varied with a controlled term u, then

$$s_2 = u + \bar{s}_2,$$

where $\bar{s}_2$ is a constant. Now the process becomes a bilinear control system. If $s_2 > 0$, energy is dissipated and the control is passive, but if $s_2 < 0$, energy must be added and the control is active. Although somewhat more difficult to synthesize, it is possible to control spring stiffness k_2 in a similar manner.

For this system it is relatively simple to see that the minimization of a quadratic-state cost function with obvious control magnitude constraints results in a bang-bang control. Although not attempted here, the STVM above could be applied to compute the optimal switching times. A linearized analysis can be used to compute the linear system solution with the switching function involving a solution to the algebraic Riccati equation as in [K13] and [K14]. However, the optimal linear solution can be improved considerably by a simple bilinear solution that maximizes convergence of the state quadratic cost by an appropriate switching function. Such control, although still suboptimal, can decrease the peak in the passenger mass frequency response by about a factor of 3 over that of conventional passive damping and almost a factor of 2 over that of the linearized derivation [K15]. Here the frequency response refers to vertical velocity and acceleration of the passenger mass. Simulation details are left for the reader.

7.3.2 Nuclear Rocket Engine Control

Control of a nuclear power reactor and other processes which combine to form a nuclear rocket propulsion system is studied here. Such rocket systems have been researched and ground tested for long-range manned space missions. They are desired, and possibly required, because of their high specific impulse (thrust to propellant mass flow rate ratio) which results in large payloads relative to that of chemical systems. The specific impulse can be shown to be nearly proportional to nozzle-entrance propellant temperature divided by the square root of the propellant molecular weight. This leads to the success of relatively fast fissioning, high-temperature reactors with hydrogen (molecular weight of 2) coolant (propellant). The control of nuclear rockets and their optimization is studied in [M2], [M5], and [M8]. Only a brief summary is presented here.

Figure 7.10 illustrates a nuclear rocket system. Although not shown here, a turbine-driven pump may be powered by bleeding hot gas from the reactor (so-called bleed-cycle process). The liquid hydrogen is pumped through the nozzle jacket and reactor reflector, where it is preheated to a gaseous state prior to entering the reactor core. In the core, the hydrogen is heated to very high temperature by neutrons from the fission process, which is controlled by neutron-reflecting vanes in the reactor reflector. The high energy of the nozzle-expanded hydrogen propellant imparts a high thrust at high specific impulse. The specific impulse typically is several times higher than that of chemical engines. The thrust is controlled by the engine propellant bleed valve, which indirectly manipulates flow rate and nozzle pressure. The latter leads to thrust for a specified temperature or specific impulse. Typical design parameters are given in [M8] and summarized in Table 7.2.

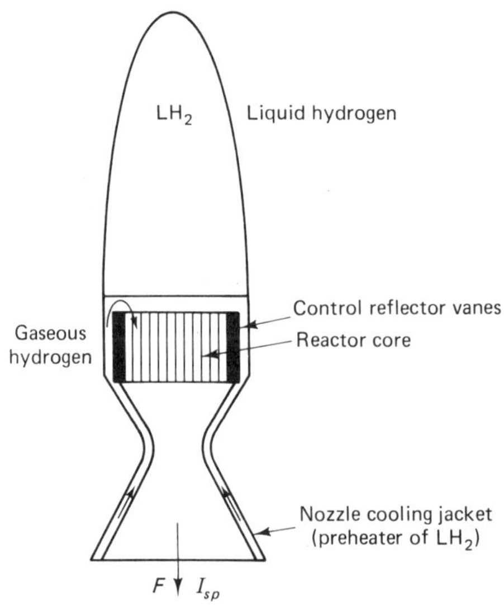

Figure 7.10 Nuclear rocket engine system. F, thrust; I_{sp}, specific impulse.

The total system involves a complex coupling of certain fast dynamics and relatively slow dynamics. The single-lump linearized heat-transfer time constant of the core at design conditions is about 1.5 sec and is much slower (being nearly inversely proportional to coolant flow rate) during startup. On the other hand, the neutron density (power level) responds in a fraction of a second. Consequently, approximate control analysis (as a first attempt) can proceed in a decentralized fashion.

TABLE 7.2 DESIGN CONDITIONS FOR HYPOTHETICAL ROCKET ENGINE

Thrust	100,000 lb
Specific impulse (including losses)	760 sec
Reactor power	2260 MW
Propellant weight flow rate	130 lb/sec
Nozzle entrance stagnation pressure	1100 psia
Reactor exit temperature	4500°R
Reflector entrance pressure	1220 psia
Nozzle exit pressure	4.0 psia
Nozzle throat area	61 in.^{2}
Nozzle expansion ratio	20
Gas temperature at reflector entrance	120°R
Gas temperature at core entrance	220°R
Neutron mean effective generation time	1.4×10^{-5} sec
Propellant reactivity	0.0280
Temperature reactivity	−0.0073
Heat exchange thermal time constant	1.5 sec
Control actuator bandwidth and damping	50 sec^{-1}, 0.7
Turbopump speed, closed-loop bandwidth	40 sec^{-1}

Source: Ref. [M8].

To maximize mission payload, it is desired to start up the system to design level so as to minimize coolant consumption. Consequently, it is necessary to keep the coolant flow rate as low as possible and to bring the reactor to design power as fast as possible within the allowable physical constraints which are placed on control and state variables. Time-optimal state control of the nuclear fission process was studied in Example 6.5.

Although the reactor state control has its utility (e.g., during calibration of instrumentation when time is valuable), it is frequently desirable to traverse from some initial equilibrium state to a terminal equilibrium neutron level in minimal time without the terminal precursor level necessarily being in equilibrium. This problem may be specified merely by defining the terminal set to be the vertical line $n = n_f$ in the n versus c state plane shown in Fig. 6.9. Here the projection of the costate vector on state space must be perpendicular to this terminal line at the terminal time t_f.

With $p_2(t_f) = 0$ and time reversed, it is seen from Fig. 6.10 that $p_1(t)$ can have no zeros for $t < t_f$. Similar plots, with the eigenvectors in the same quadrants, could be obtained for other values of γ. Hence the time-optimal control with a reactivity constraint is constant for $t_0 < t < t_f$. That this process requires no switching on (t_0, t_f) may also be shown by the solution to the adjoint system with time reversed, which is similar to the neutronics solution.

At t_f, the terminal time, the neutron precursor level is not yet at equilibrium. Therefore, reactivity must be different from zero to maintain constant neutron level

$n(t) = n_f$ for $t > t_f$. From (6.74) it is seen that

$$u = \frac{l\dot{c}}{n_f} = \beta - \frac{\lambda l c(t)}{n_f} \tag{7.42}$$

is the required terminal reactivity. From (6.81) it is readily shown that this required terminal control does belong to the admissable set with $|u(t)| \le \gamma$.

If the switching function of the control equation vanishes for any finite time, the control is said to be singular. For (6.80) a singular control exists on some time interval if $p_1(t)$ is zero on this time interval. It is seen from (6.79), however, that this would require $p_2(t) = 0$. Hence $\mathbf{p}(t) = \mathbf{0}$ on this finite interval, which would contradict the maximum principle. Therefore, the problem does not have a singular solution. Along with uniqueness of the bang-bang-controlled state trajectories, it is obvious that the extremal neutronic policy derived here is indeed the unique time-optimal control.

If six precursor groups are considered, the state is described by (2.120) and (2.121) in Vol. 1. The time-optimal control is still obtained from (6.80), but in this case the switching function $p_1(t)$ satisfies the adjoint equations of (2.120) and (2.121), Vol. 1. Again, the problem of most physical interest is to control the fission process from an initial point to a terminal hyperplane, $n = n_f$, in a seven-dimensional state space. The transversality condition then requires that

$$p_i(t_f) = 0 \qquad \text{for } i \neq 1.$$

From this condition it is readily seen from the homogeneity of the costate equations that $p_1(t)$ cannot have a zero in finite time. Consequently, $u(t)$ does not switch on (t_0, t_f).

Again, the terminal control for $t > t_f$ must compensate for the decaying precursors. In this case, the terminal control is

$$u = \frac{l}{n_f} \sum_{i=1}^{6} \frac{dc_i}{dt}. \tag{7.43}$$

It may be seen from the state equations that this terminal control is admissible similar to the one-precursor case.

It is shown in [M8] that a small error in measurement to estimate the required terminal control results in a divergence of $n(t)$ away from n_f. A sliding control that holds neutron level essentially constant has been synthesized. The idea is to allow $n(t)$ to vary within certain small prescribed limits about n_f, and to drive the control at maximum effort about the theoretical in order to correct $n(t)$ when its deviation exceeds these limits. Figure 7.11 shows a rapid time-optimal start controlled at the terminal end by such a sliding (actually chattering) process. Here, reactor power (proportional to neutron level) is held essentially constant after a 9-msec increase by

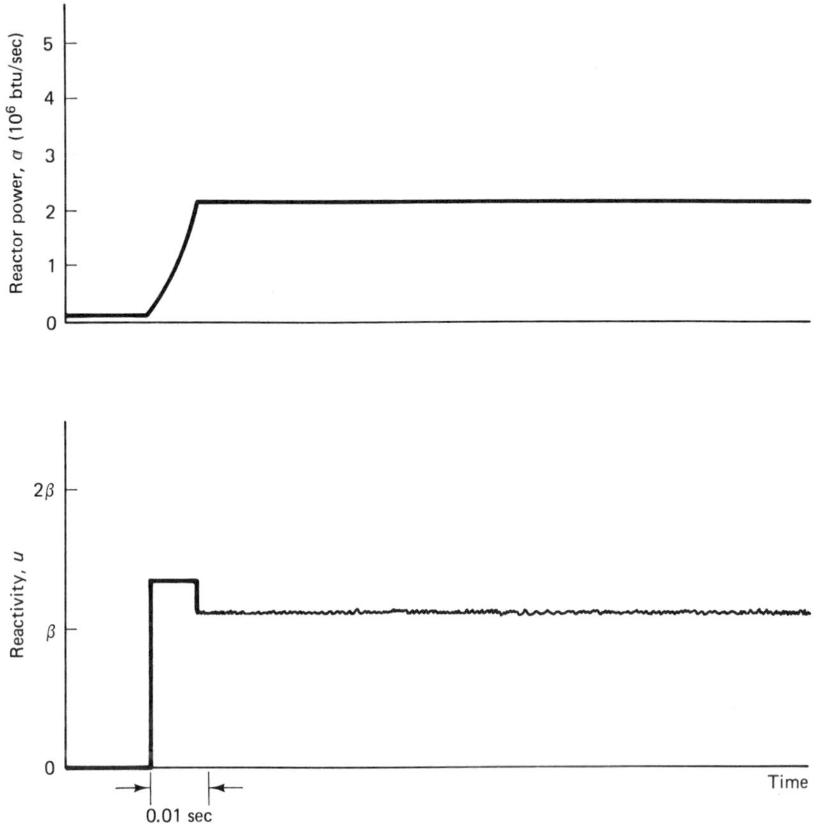

Figure 7.11 Time-optimal start with sliding (chatter) terminal control. (From [M8].)

a factor of 20 in power. The key to the synthesis of the time-optimal neutron level control appears to be the ability to maintain neutron density essentially constant while the precursor level is not near steady state.

Although the bang-bang neutronic control is synthesized very simply, the required terminal control is slightly more complicated. In many cases a conventional type of closed-loop control is satisfactory and may even approach the performance of the optimal process. Even for these cases, however, the optimization analysis provides a yardstick of performance. One convenient technique of approximating the optimal process is to synthesize a continuous feedback control system which is fast-acting but limited in control variations. The startup of such a system is shown in Fig. 7.12. Initially, the system is at equilibrium, with a neutron level of less than one-twentieth of the terminal level. Then 0.9β of reactivity is added until the neutron level is about one-eighth of the terminal value. At that time, a proportional-plus-integral type of feedback controller replaces the constant control process. The con-

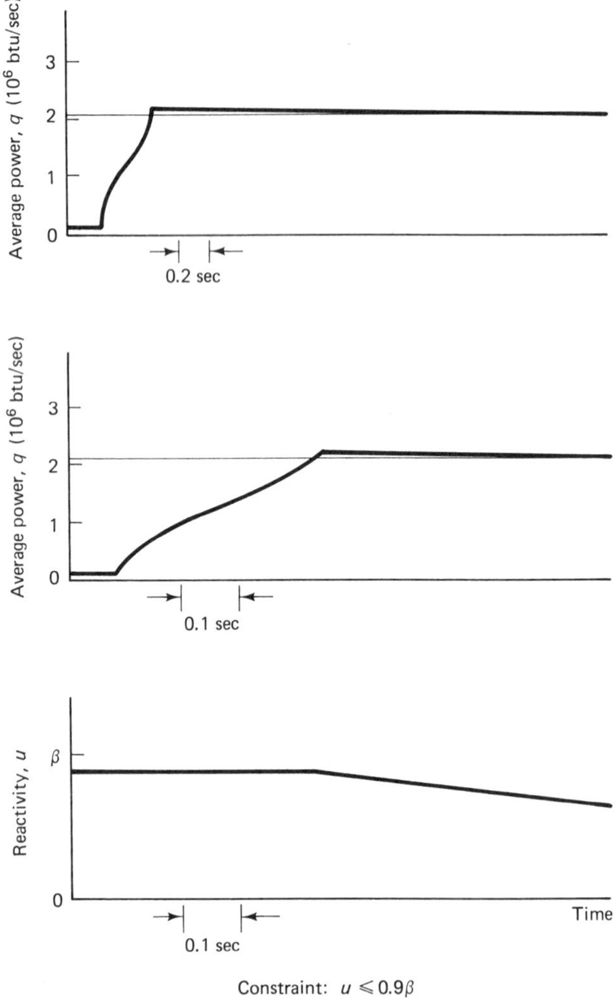

Figure 7.12 Time-optimal neutronic start with continuous terminal control. (From [M8].)

troller was introduced at that time to limit the magnitude of the controller integral signal. (Log-power control or a separate integral controller limit would allow the loop to be closed for the entire startup.) The power controller used here is defined by

$$u = 1.87(10)^{-9}q_e + 6.24(10)^{-9}\int_{t_0}^{t} q_e \, d\sigma, \tag{7.44}$$

where $q_e(t) = q_f - q(t)$ (Btu/sec) and $q \propto n$. Due to the high gain of the system, the reactivity stays at its constrained value of 0.9β until the terminal state is approached.

After a small rise time, $q(t)$ has a small overshoot followed by a slow decay toward the desired steady-state value. The slow decay is caused by the slow birth of precursor neutrons. For most applications, such small overshot is insignificant, and this simple suboptimal closed-loop control could be utilized. Nevertheless, the sliding terminal control maintains constant power more accurately.

For many nuclear fission processes, it is possible to change reactivity quite rapidly, and for these, inertia-less control such as that assumed in the preceding sections is a good assumption. In other cases, however, control may be slow-acting due to physical inertia and energy constraints. Sometimes, reactivity is even velocity limited as well as magnitude limited for safety reasons. To apply the maximum principle (with state constraints) to these problems, it is convenient to define reactivity as a new state variable, say $x_3(t)$, for a single-precursor model. Then let the rate of change of reactivity be the control $u(t)$. Now the neutron kinetics are described by

$$\frac{dx_1}{dt} = \frac{x_3 - \beta}{l} x_1 + \lambda x_2,$$

$$\frac{dx_1}{dt} = \frac{\beta}{l} x_1 - \lambda x_2,$$

and

$$\frac{dx_3}{dt} = u. \tag{7.45}$$

In terms of previous notation, the state variables are $x_1 = n(t)$ or $q(t)$, $x_2 = c(t)$, and $x_3 = \delta k(t)$, reactivity. The dynamics are constrained so that (for mathematical convenience and physical limitation)

$$g(\mathbf{x}) = x_3^2 - \gamma^2 \leq 0 \tag{7.46}$$

becomes a state constraint, and

$$|u(t)| \leq \eta \tag{7.47}$$

is a control constraint.

For the first problem find the admissible time-optimal reactivity variation to control the neutron kinetics from an initial equilibrium state $\mathbf{x}(t_0) = \mathbf{x}_0$ (with $x_{20} = \beta x_{10}/l\lambda$ and $x_{30} = 0$) to a desired terminal equilibrium state $\mathbf{x}(t_f) = \mathbf{x}_f$ (with $x_{2f} = x_{1f}/l\lambda$ and $x_{3f} = 0$). Again,

$$u = \eta \operatorname{sgn} p_3(t) \tag{7.48}$$

is the form of the control that makes

$$\mathcal{H}(\mathbf{x}, \mathbf{p}, u) = \mathbf{p}^{\mathrm{T}}\mathbf{f}(\mathbf{x}, \mathbf{p}, u) = p_1\left[\left(x_3 - \frac{\beta}{l}\right)x_1 + \lambda x_2\right] + p_3 u + p_2\left(\frac{\beta x_1}{l} - \lambda x_2\right)$$

a maximum and is a candidate for the time-optimal strategy while $|x_3(t)| < \gamma$.

While the solution to the maximum principle $\mathbf{x}(t)$ remains interior to the state constraint boundary, $g(\mathbf{x}) = 0$ as specified here, the maximum principle without state constraints remains valid. At the entrance corner to this boundary, however, the costate may have a jump discontinuity, and while $\mathbf{x}(t) \in \partial g$ [i.e., $\mathbf{x}(t)$ is the boundary $g(\mathbf{x}) = 0$], it can be shown from constrained first variations [M8] that

$$\frac{d\mathbf{p}}{dt} = -\left[\frac{\partial \mathbf{f}}{\partial \mathbf{x}} - \frac{\partial \mathbf{f}}{\partial u}\left(\frac{\partial \mathbf{x}}{\partial u}\right)^{-1}\left(\frac{\partial z}{\partial \mathbf{x}}\right)^{\mathrm{T}}\right]^{\mathrm{T}} \mathbf{p}, \tag{7.49}$$

where $z(\mathbf{x}, u) = dg/dt = (\partial g/\partial \mathbf{x})^{\mathrm{T}}\dot{\mathbf{x}}$. For (7.46),

$$z = 2x_3\dot{x}_3 = 2ux_3$$

or while $g(\mathbf{x}) = 0$, $u = 0$ and hence $z = 0$.

Consequently, for all $t \in (t_0, t_f)$, the switching function is determined by $\dot{\mathbf{p}} = -\partial\mathcal{H}/\partial\mathbf{x}$, or

$$\begin{aligned} \frac{dp_1}{dt} &= \left(\frac{\beta}{l} - x_3\right)p_1 - \left(\frac{\beta}{l}\right)p_2, \\ \frac{dp_2}{dt} &= \lambda(p_2 - p_1), \\ \frac{dp_3}{dt} &= -x_1 p_1. \end{aligned} \tag{7.50}$$

It is apparent that a singular control $p_3(t) = 0$ in (7.48) cannot satisfy the maximum principle, for, from (7.50), $p_3 = 0$ requires that $p_1 = p_2 = 0$. Thus, since the trajectories resulting from the maximum principle are unique, (7.48) is the optimal control if $|x_3| < \gamma$, and $u = 0$ while $|x_3(t)| = \gamma$.

It is shown in [M8] that the switching function $p_3(t)$ can have at most four zeros as exemplified in Fig. 7.13. In practice, the negative reactivity portion is relatively smaller than shown here and usually can be neglected, since the resulting terminal error is smaller than instrumentation error.

Again, it may be desired to control the kinetics from some initial equilibrium state $\mathbf{x}_0$ to some $x_1(t_f) = x_{1f}$, so that $|x_3(t)| \leq \gamma$, $|u(t)| \leq \eta$ for all t, and $dx_1(t)/dt = 0$ for $t \geq t_f$. From the terminal condition, $p_2(t_f) = 0$, it can be seen

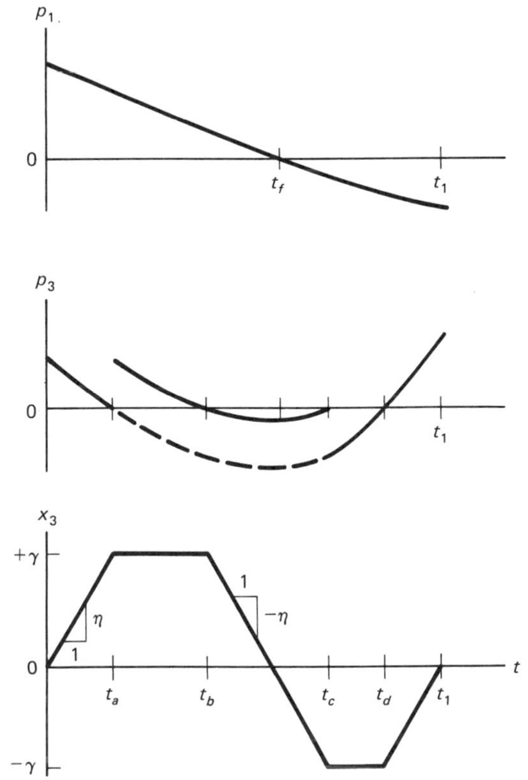

Figure 7.13 Time-optimal reactivity and switching function.

from (7.50) that a continuous $p_3(t)$ can have at most one zero on $[t_0, t_f]$ since $p_1(t)$ cannot have any zeros for this case (similar to the preceding problem). Consequently, with a jump at the state constraint entering corner, $p_3(t)$ cannot have more than two zeros, and the time-optimal reactivity variation is a truncated trapezoid on $[t_0, t_f]$. Here t_f is much smaller than for the previous problems. Such inputs and corresponding responses are shown in Fig. 7.14. It can be seen from (7.45) that $x_3(t)$ is continuous at the terminal end as shown in Fig. 7.14. It is proven in [M8] that adding more precursors to the model does not alter the maximum number of switchings of the control.

Now consider the heat-exchange process from Vol. 1, (2.124)–(2.126), which in approximate form is given by (2.128). That is, core temperature change is

$$\frac{dT_u}{dt} = \frac{q}{c_m} - awT_u, \tag{7.51}$$

where a is a heat transfer constant, c_m thermal capacity and w is coolant flow rate.

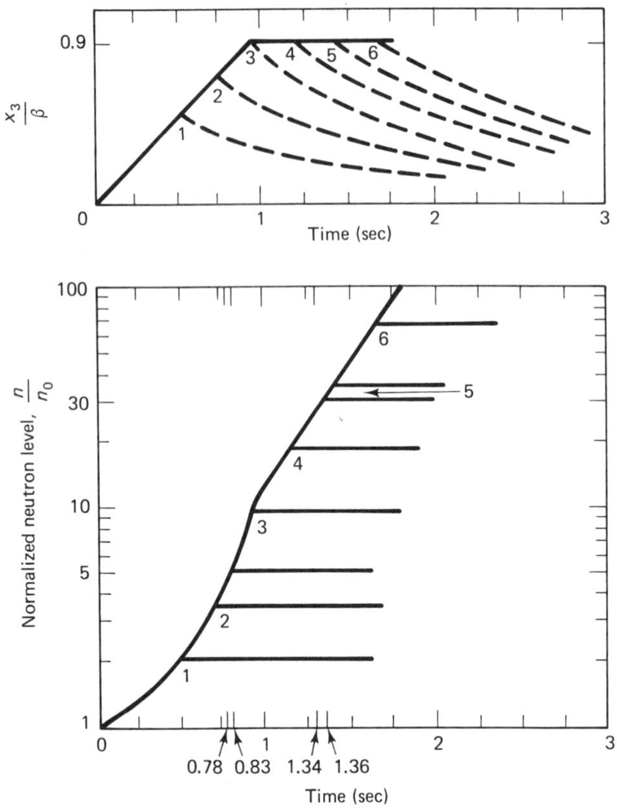

Figure 7.14 Time-optimal neutronic control with reactivity rate constraint.

Suppose that it is assumed for the present that $q(t)$ and $w(t)$ are at least piecewise-continuous controls with upper and lower physical constraints so that

$$0 \leq q(t) \leq q_b$$

and

$$w_a \leq w \leq w_b.$$

If it is desired to raise the average core temperature in minimum time or with minimum coolant, it is obvious from the maximum principle or physical reasoning that

$$q(t) = q_b$$

and that

$$w(t) = w_a \qquad \text{on } (t_0, t_f).$$

At the endpoints, $T_u(t_0) = T_{u0}$ and $T_u(t_f) = T_{uf}$ with $dT_u/dt = 0$. Therefore,

$$q(t_0) = ac_mT_{u0}w(t_0) \qquad \text{and} \qquad q(t_f) = ac_mT_{uf}w(t_f).$$

The reactor control follows Vol. 1, Section 2.3.3. Temperature reactivity is given by Vol. 1, (2.129), or

$$\delta k_t = \alpha_t T_u, \tag{7.52}$$

where α_t generally is a negative constant. The hydrogen propellant used with the nuclear rocket is a good moderator. Even as a dense gas at the reactor core entrance, it has a strong moderating effect on intermediate energy (thermal) neutrons and on fast neutrons. Such coolant reactivity is approximated by

$$\delta k_p = k_p \rho_g V, \tag{7.53}$$

where k_p is a positive constant, V the coolant volume in the core, and ρ_g the coolant density. Although the reactivity feedback is positive, it is apparent that again the temperature effect is negative according to the gas equation,

$$\rho_g = \frac{p}{zRT_g}, \tag{7.54}$$

where p is the average coolant pressure in the core, R the universal gas constant, and z the compressibility factor (usually of value 1). The coolant density reactivity also couples the neutron kinetics to the coolant feed system.

The total reactivity δk, which is the input to the neutron kinetics, becomes

$$\delta k = \delta k_t + \delta k_p + \delta k_c, \tag{7.55}$$

where δk_c is the reactivity generated by the neutron reflectors in the form of control vanes. Since meaningful safety constraints are put on δk rather than δk_c, it is assumed here that the admissible control must satisfy $|\delta k| \leq \gamma$. (Symmetrical constraints are used for convenience.) This has the effect of decoupling the control loops, as is analyzed next.

Suppose that the reactor dynamics are described by the nuclear fission process, (6.24), with $q(t)$ proportional to $n(t)$, and by the simple heat exchange process (7.51). Let the control variables be $u_1 = \delta k(t)$ and $u_2 = w(t)$, and combine constants in (7.53) and (7.54), so that

$$k_p = \frac{c_p u_2}{(T_g)^{1/2}}, \tag{7.56}$$

where c_p is a positive constant.

Now the problem is to start the system on minimal coolant ($f_0 = u_2$) from some initial equilibrium state with $q(t_0) = q_0$, $T_u(t_0) = T_{u0}$, and $u_2(t_0) = u_a$ to some terminal condition so that $q(t_f) = q_f$, $T_u(t_f) = T_{uf}$. Also, $\dot{q}(t) = 0$ and $\dot{T}_u(t) = 0$ for $t > t_f$. Control and state constraints include

$$|u_1(t)| \leq \gamma, \tag{7.57}$$

$$w_a \leq u_2(t) \leq w_b, \tag{7.58}$$

and

$$q(t) \leq q_f. \tag{7.59}$$

Reasons for the control constraints have been discussed, and the state constraint on power may be a maximum attainable power for the critical assembly, a safety constraint on power overshoot, or a temperature time rate of change constraint at a fixed flow rate.

It is apparent from the maximum principle that the optimal control while $\mathbf{x}(t)$ is not on a state constraint boundary is described again by a bang-bang process with

$$u_1 = \operatorname{sgn} p_1(t), \tag{7.60}$$

$$u_2 = \begin{cases} w_a & \text{for } S(\mathbf{x}, \mathbf{p}) < 0, \\ w_b & \text{for } S(\mathbf{x}, \mathbf{p}) > 0, \end{cases} \tag{7.61}$$

where

$$S(\mathbf{x}, \mathbf{p}) = -1 - ap_8T_u \tag{7.62}$$

and $\mathbf{p}$ is described again by

$$\frac{d\mathbf{p}}{dt} = \frac{-\partial H}{\partial \mathbf{x}}, \tag{7.63}$$

with $H = -f_0 + \mathbf{p}^{\mathrm{T}}\mathbf{f}$. Since there is a constraint on total reactivity, $u_1(t)$, no increase in coolant flow rate, $u_2(t)$, would permit a faster change of state nor a consumption of less coolant than does $u_1(t) = \gamma$ and $u_2(t) = w_a$. At the power constraint boundary, however, it is again necessary that

$$u_1 = l \sum_{i=1}^{6} \frac{\dot{c}_i}{q_f}.$$

Again, this control is unstable in practice, but may be approximated by a simple closed-loop control or a sliding control. At the desired terminal temperature, the flow rate is increased so that $u_2(t_f) = q_f/c_m a T_{uf}$. For the nuclear rocket, this corresponds to the designed thrust and temperature condition. Figure 7.15 shows the

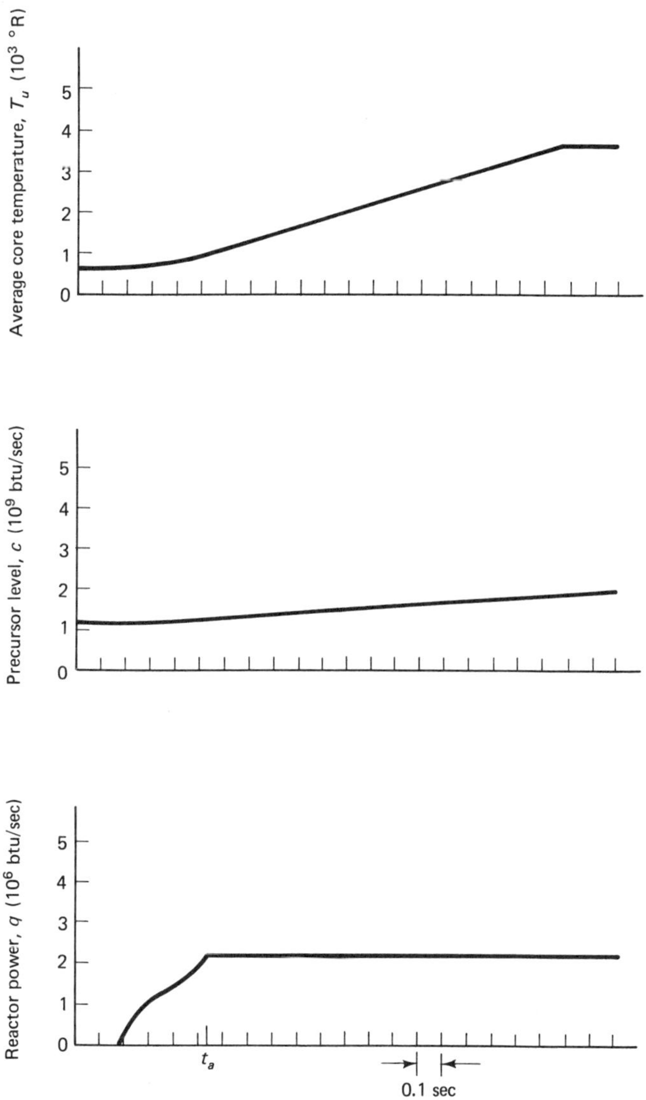

Figure 7.15 Minimal propellant control of a nuclear rocket. (From [M8].)

minimal propellant control of a rocket reactor whose design parameters are given in Tables 2.1 (Vol. 1) and 7.2. It may readily be seen that these trajectories satisfy the maximum principle and further are indeed optimal trajectories [M8].

Table 7.3 quantifies pertinent state and costate values during the coolant-optimal reactor start shown in Fig. 7.15. The costate equations are defined by (7.63), while $q(t) < q_f$ and $t_0 \le t < t_a$. On the interval (t_a, t_f), $q(t) = q_f$ and the

TABLE 7.3 OPTIMAL REACTOR STARTUP TRAJECTORY WITH $q(t) \le q_f$ [a]

Time, t (sec)	Power, q (10^4 Btu/sec)	Temperature, T_e (10^2 °R)	$p_1(t)$ (10^{-6} sec^2/Btu)	$p_8(t)$ (10^{-4} sec/°R)
0	1.3144	5.000		
$t_0 = 0.1$	1.3144	5.0000	11.946	7.2139
0.2	13.253	5.0651	9.2838	7.2306
0.4	24.889	5.33720	5.6093	7.2640
0.6	24.889	5.9257	3.3932	7.2976
0.8	71.793	6.8790	2.0589	7.3313
1.0	119.91	8.4903	1.2663	7.3652
$t_a = 1.24$	214.20	11.911	5.0718	7.4101
1.4	214.20	14.868	4.9676	7.4375
1.6	214.20	18.549	4.8368	7.4718
1.8	214.20	22.213	4.7054	7.5064
2.0	214.20	25.860	4.5734	7.5411
2.2	214.20	29.490	4.4408	7.5759
2.4	214.20	36.701	4.3076	7.6110
2.6	214.20	36.701	4.1737	7.6461
2.8	214.20	40.281	4.0393	7.6815
$t_f = 3.08$	214.20	45.011	3.8500	7.7751

Source: Ref. [M8].

[a] Input data: $u_1 = 0$, $t < 0.1$ sec $= t_0$; $u_1 = 0.9\beta$, $t \ge 0.1$, $p_{10} > 0$, and $q < q_f = 2.142(10)^6$ Btu/sec, for $t < t_a$; $u_1 = l\Sigma_{i=1}^{6}\, \dot{c}_i/q_f$, for $t > t_a$; $u_2 = u_a = 2$ lb/sec, $T_u < T_{uf}$, and $S < 0$, for $t > t_f$; $a = 1.153 \times 10^{-2}$, $T_u(0) = 500$°R; $dT_u(t_1)/dt = dq(t_f)/dt = 0$. Other data given in Tables 2.1 and 7.2, Vol. 1.

costate is defined by (7.49) or

$$\frac{d\mathbf{p}}{dt} = \frac{-\partial H}{\partial \mathbf{x}} + \rho\left(\frac{\partial z}{\partial \mathbf{x}}\right), \tag{7.64}$$

where $z(\mathbf{x}, \mathbf{u}) = \dot{q}(\mathbf{x}, \mathbf{u})$. Also, with $u_2(t) = w_a$ on (t_a, t_f), it is necessary from the addended maximum principle [P4] that

$$\frac{\partial H}{\partial \mathbf{u}} = \rho \frac{\partial z}{\partial \mathbf{u}} + \gamma \frac{\partial r}{\partial \mathbf{u}}, \tag{7.65}$$

where

$$r = u_a - u_2. \tag{7.66}$$

These may be derived formally from first variations with added Lagrange multipliers for both the state and control constraints taken simultaneously. Further, it is seen that

$$\gamma = S(t) = -1 - ap_8T_u$$

and

$$\rho = p_1(t).$$

Since $T_u(t_f) = T_{uf}$ and $q(t_f) = q_f$, the transversability condition requires that $p_j(t_f) = 0, j = 2, \ldots, 7$.

At the constraint entrance corner, that is, when $t = t_a$, $\mathbf{p}(t)$ may be assumed continuous since there is no exit corner [P4]. Again, it is readily seen that the problem cannot have a nondegenerate singular solution.

The preliminary analysis presented here for the nuclear rocket control is typical of that required for complex nonlinear systems in engineering practice. Of course, there are many stages in the research, design, testing, and development of such a system. Again, the importance of integrating system design with dynamic and control analysis cannot be overemphasized. One of the most significant contributions of an optimal control study is the contribution to system design that pertains to sensitivity of performance with respect to plant design parameters. The above is a good example. Like most high-performance adaptive systems, the process frequently operates on a control or state constraint boundary. Reactor exit coolant temperature is an obvious state variable that directly affects engine specific impulse (thrust/coolant mass flow). Consequently, it is crucial that the coolant (propellant) be heated to as high a temperature as possible and that the reactor core be designed with appropriate structure and material to withstand such high temperature. Thermal stress constraints, which are related to both spatial and temporal temperature gradients, can be significant. Although not considered above, they may be implemented in an approximate manner by appropriately scheduling less reactor power during a minimal propellant start when the temperature gradient constraint is reached. At the same time, coolant flow rate from the pump is kept as low as possible. Unfortunately, there is a critical stall region in the pump performance map which is shown in Fig. 7.16. The stall line is defined approximately by a line of so-called pump specific speed. When the coolant temperature is increased to a certain value while there exists a minimum coolant flow rate w_a, it becomes necessary to schedule an increase in the flow rate to an appropriate value to avoid pump stall, as shown in Fig. 7.16. Again significant savings in coolant, tank weight, and so on, can result in increased payload for the mission.

A series of analyses, simulations, and tests of this sort leads eventually to a satisfactory design configuration. The final system may involve a distribution of microprocessors to perform on-line estimation of critical states and parameters, and to adapt controller policies to changing conditions. In any event, the foregoing type of optimization analysis, as important as it may be, does not provide the final control synthesis.

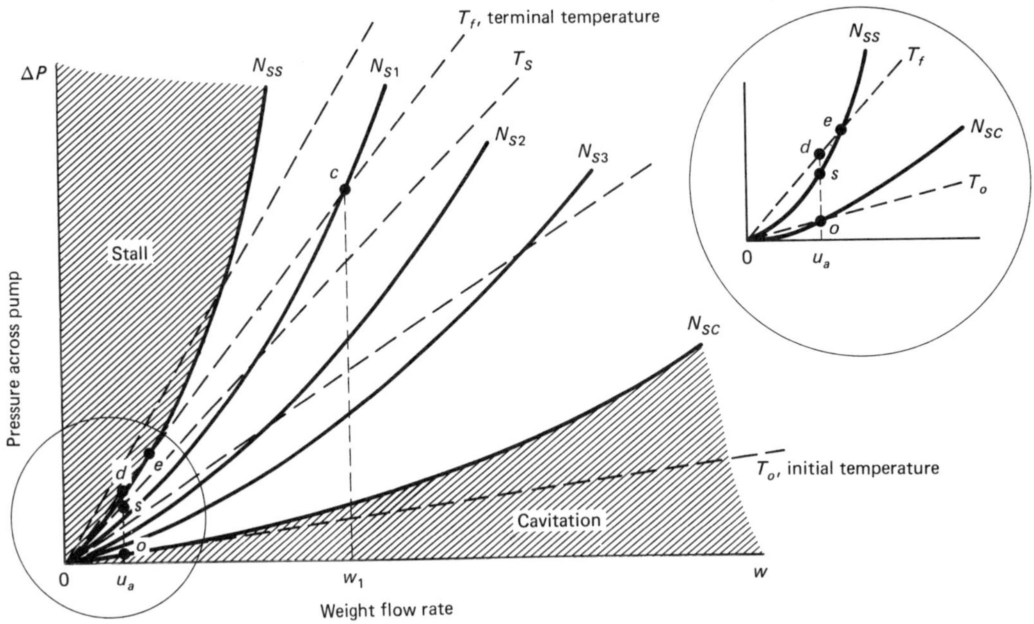

Figure 7.16 Hypothetical pump performance map. *Note:* Temperatures refer to the nozzle entrance and are approximately linear. The specific speed is $N_s = NQ^{1/2}/H^{3/4}$, where N is the pump speed (rpm), Q the flow rate (gpm), and H the pump pressure head (feet). (From [M8].)

7.3.3 Adaptive Control Application

The previous nonlinear control suggests a high-performarnce adaptive system with on-line estimation and control adjustment. One interesting method of rapid identification and adaptive control utilizes the cross-correlation of a system output with an approximate white noise input. Recall from the study of random processes that the autocorrelation of an ergodic random process $n(t)$ is given by [P6]

$$R_n(\tau) = \lim_{T\to\infty} \frac{1}{2T} \int_{-T}^{T} n(t - \tau)n(t)\, dt. \tag{7.67}$$

Similarly, the cross-correlation of two jointly ergodic random processes $n(t)$ and $x(t)$ may be expressed as

$$R_{nx}(\tau) = \lim_{T\to\infty} \frac{1}{2T} \int_{-T}^{T} n(t - \tau)x(t)\, dt. \tag{7.68}$$

If $n(t)$ is the input to a linear system of impulse response $h_1(t)$, the convolution integral relates the output $x(t)$ to the input, which after substitution of (7.67) and (7.68) yields

$$R_{nx}(\tau) = \lim_{T\to\infty} \frac{1}{2T} \int_{-T}^{T} h_1(\sigma) R_n(\tau - \sigma)\, d\sigma. \tag{7.69}$$

If $n(t)$ is a white-noise process,

$$R_n(\tau) = a\delta(\tau), \tag{7.70}$$

where a is a positive constant signal power. Then, from the convolution integral (7.69),

$$R_{nx}(\tau) = ah_1(\tau). \tag{7.71}$$

It is quite easy to implement such a cross-correlation computation with very small binary pseudorandom sequences for $n(t)$ by means of a microprocessor.

For a nonlinear system with changing equilibrium conditions of operation, the operation can be performed for on-line dynamic linearized system identification. Then controller gains can be adjusted according to some desired linear-system response about the equilibrium of interest. The amplitude of excitation signal can be so small as not to appreciably disturb system operation. Also, it can be shown that the main signal input usually is uncorrelated with the noise excitation. Consequently, the main signal only results in a constant term in the cross-correlation operation which can be subtracted off. A block diagram for such an adaptive control of the so-called longitudinal aircraft dynamics is shown in Fig. 7.17 [A5].

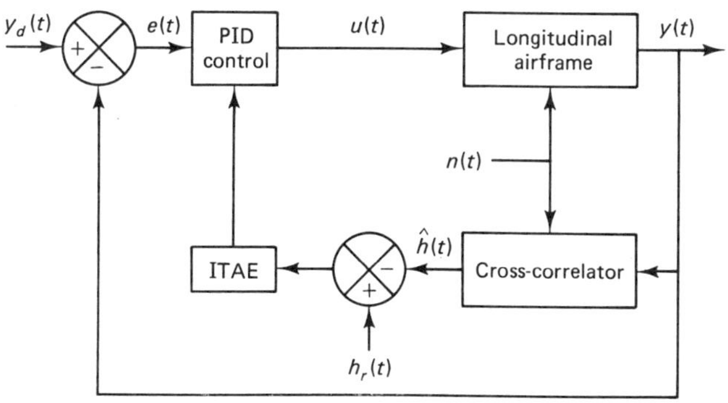

Figure 7.17 Self-optimizing autopilot.

For the longitudinal-aircraft dynamics, the linearized transfer functions at given Mach number, altitude, and aircraft trim conditions results in two second-order modes. The short-period mode typically has higher damping than the so-called phugoid mode, which can be very lowly damped but of slow oscillation. Pitch motion exhibits primarily the short-period mode, and airspeed changes predominantly exhibit the phugoid oscillation. Both natural frequencies and damping coefficients can vary drastically with different flight conditions (Mach and altitude). In fact, for modern, large-angle-of-attack manuevering aircraft (poststall flight), the short-period damping can be negative (unstable). These conditions almost dictate adaptive control such that the final synthesis is nonlinear and typically bilinear. Such control may appropriately adjust optimal linear regulators—the latter as designed for the Boeing 767 [G5].

The purpose of the adaptive controller might be to adjust the proportional-plus-integral-plus derivative (PID) controller gains to maintain a desired response characteristic. It has been found that the integrated-time-absolute value of the error (ITAE) criterion is a good indication of relative damping characteristic. Therefore, the ITAE controller adjustment is used in Fig. 7.17. The relative success of such an

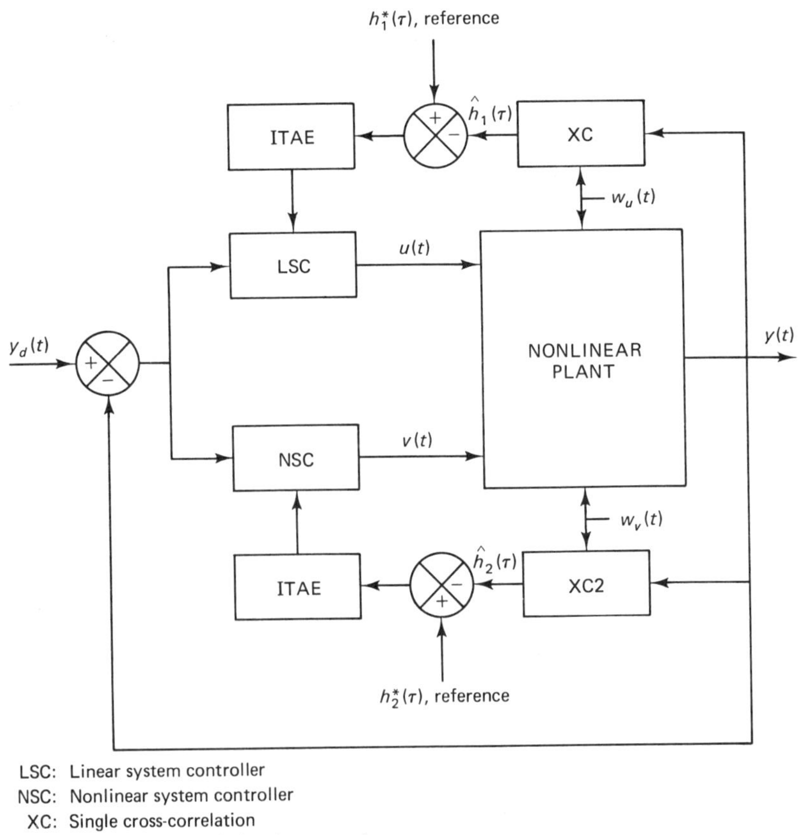

LSC: Linear system controller
NSC: Nonlinear system controller
XC: Single cross-correlation
XC2: Double cross-correlation ($\tau = \tau_1 = \tau_2$)
ITAE: Integrated time absolute-value of error (or other) min. adjustment

Figure 7.18 Nonlinear self-adaptive controller.

adaptive controller was demonstrated for aircraft control of pitch motion with input on elevator control angle, even before the advent of fast microprocessors.

The method also has been used to characterize the linearized dynamics of a nuclear rocket reactor [B15]. This is extended to the nonlinear system's second-order kernel in [B16]. The latter utilizes a pseudorandom ternary sequence for the small random input on reactivity. The second-order time-averaged cross-correlation of output power with the small reactivity variation provides an estimate of the second-order Volterra kernel. Figure 7.18 shows an adaptive nonlinear control scheme based on this principle as an extension of the adaptive linear controller of Fig. 7.17. Other investigators suggest the use of the Wiener kernels [S5], which have the added convenience of orthogonality. It is anticipated that such methods eventually will be useful for nonlinear systems adaptive control—an area of surprisingly limited research despite its great potential.

7.4 CONCLUDING REMARKS AND DIRECTIONS

The culmination of nonlinear system dynamic analysis into preliminary system design is presented here. Applications are so complex by their nature that only a limited survey could be presented in the allotted space. While variable-structure systems and nonlinear systems in general are adaptive in nature, only limited results have been obtained in this direction. Adaptive linear-system control, on the other hand, has been a very active area of research and may provide a clue to effective utilization of adaptive nonlinear characteristics. Among others, excellent references in this area include [A6]–[A8], [K16], [N1], [S6], and [T3]. It is noted that the typical linear-system adaptive control by parametric adjustment is really a bilinear control system, and much of the theory presented here should be useful in this respect. As noted in Section 7.3.3, aircraft control represents an excellent application for adaptive control. The introduction, which is presented here, is only a sample of possible approaches. Again, additive linear control seems appropriate for local stability and controllability with parametric control useful for large variations of state.

Robotics and automated manufacturing are very nonlinear as demonstrated in Vol. 1, Chapter 2. Yet most of the control work in this area has involved linearized trajectory following policies. Reference [F3] compares several interesting robotic control systems, including adaptive systems. An adaptive, self-tuning controller with an autoregressive model to fit input–output manipulator data is presented in [K17]. Reference [L8] considers joint interactions, which were omitted in [K17], to study an adaptive, linearized, trajectory-following controller for a robot manipulator. A least-squares performance index, which is a special case of the general quadratic cost introduced in Section 6.2.2, is used to compute the optimal policy.

Sliding control was shown to be effective for the nuclear reactor control in Section 7.3.2. More generally, this can be a very convenient method of nonlinear system control with performance independent of plant dynamics for the most part.

Singular surfaces and state constraints sometimes can be accurately realized by such sliding or chattering control. A good survey of this topic and variable-structure systems is given in [D7] and [A9].

It is obvious that the human body and nature in general utilize adaptive control, which probably is optimal in some sense. Several such nonlinear processes were modeled in Chapter 2. The field of optimal therapy, drug dosage, design or adaptive artificial organs, and general health care is ripe for research and development. Thus far, however, such contributions have been few, but active basic research has recently evolved in this area.

In conclusion, the world is comprised of nonlinear systems, and nonlinear engineering designs can lead to superior performance. Unfortunately, the problems are difficult, and limited tools of analysis are available. Continued research and the development of computer-aided design should be fruitful endeavors to broaden the scope of methodologies and applications.

EXERCISES

7.1. Consider the launch of a point unit mass (rocket) in the atmosphere as described by $\ddot{x} + c_d\dot{x}^2 + g = u$, where c_d (aerodynamic drag coefficient) and g (gravitational acceleration) are both positive. The thrust u is constrained by $0 \le u \le \alpha$ and $\int_0^{t_f} u(t)\,dt = \beta$. Compute the optimal $u(t)$ and $x(t)$ to maximize the terminal velocity $\dot{x}(t_f)$.

7.2. For the inertialess telescopic arm of Exercise 2.7, Vol. 1, find the optimal control u, w to move from x_0,y_0 to x_f,y_f in time t_f while minimizing

$$J = \int_0^{t_f} rw^2\,dt,$$

where $r^2 = x^2 + y^2$. This example assumes a single link with rotation ψ about the z axis, and controls $w = \dot{\psi}$ and $u = \dot{a}$.

7.3. For the two-link robot of Vol. 1, Fig. 2.9, studied in Section 2.2.3, simplify the equations of motion by neglecting the inertia of the second link ($m_2 = 0$). The controls are the torque τ_1 and u such that $\dot{\psi}_2 = u$, $\tau_1 \le T_1$, $u \le U$. Compute the optimal controls to transfer the work point from ψ_{10},ψ_{20} to ψ_{1f},ψ_{2f} while minimizing

$$J = \int_0^{\infty} (\tau_1^2 + u^2)\,dt.$$

Here motion is confined to the x-y plane; m is the mass of the first arm (at the base) and both arms have length l.

7.4. Discuss the utility and the limitations of optimal control theory in practical engineering design.

7.5. Show that the switching-time-variation method (STVM, Section 7.2) yields a solution that converges to a maximum-principle solution.

APPENDIX

A

ON LIE ALGEBRA

A.1 INTRODUCTION

A few basic concepts are provided here which are utilized in Chapters 4 and 5. Although the results can be obtained without Lie algebra, it undoubtedly would be more cumbersome. Indeed, Lie algebra seems to be particularly attractive for bilinear system analysis. First, consider the Lie bracket of two $n \times n$ matrices $\mathbf{A}$ and $\mathbf{N}$ which is given by

$$[\mathbf{A}, \mathbf{N}] \equiv \mathbf{AN} - \mathbf{NA}. \tag{A.1}$$

Properties of the Lie bracket

(i) $[\mathbf{A}, \mathbf{N}] = -[\mathbf{N}, \mathbf{A}]$.

(ii) If α and β are scalars, then

$$[\alpha\mathbf{A}, \mathbf{N}] = \alpha[\mathbf{A}, \mathbf{N}],$$

$$[\mathbf{A}, \beta\mathbf{N}] = \beta[\mathbf{A}, \mathbf{N}].$$

Notation: Let $L_A^0(\mathbf{N}) = \mathbf{N}$, and (A.2)

$$L_A^n(\mathbf{N}) = [\mathbf{A}, L_A^{n-1}(\mathbf{N})], \qquad n = 1, 2, \ldots,$$

so that

$$L_A^1(\mathbf{N}) = [\mathbf{A}, \mathbf{N}],$$
$$L_A^2(\mathbf{N}) = [\mathbf{A}, [\mathbf{A}, \mathbf{N}]]$$
$$L_A^3(\mathbf{N}) = [\mathbf{A}, [\mathbf{A}, [\mathbf{A}, \mathbf{N}]]],$$
$$L_A^4(\mathbf{N}) = [\mathbf{A}, [\mathbf{A}, [\mathbf{A}, [\mathbf{A}, \mathbf{N}]]]], \ldots .$$

(iii) Then $L_A^0(\mathbf{N}), L_A^1(\mathbf{N}), L_A^2(\mathbf{N}), \ldots$ consist of a linear algebra, with dimension no more than n^2.

Now it is possible to develop two useful algebraic relationships from this notation.

Proposition 1

$$\mathbf{A}^n \mathbf{N} = \sum_{i=0}^{n} \binom{n}{i} L_A^i(\mathbf{N}) \mathbf{A}^{n-1}. \tag{A.3}$$

Proof. From (A.2), when $n = 0$, $\mathbf{N} = \mathbf{N}$. If $n = 1$, $\mathbf{AN} = L_A^0(\mathbf{N})\mathbf{A} + L_A^1(\mathbf{N}) = \mathbf{NA} + [\mathbf{A}, \mathbf{N}]$; and by definition,

$$\mathbf{A} L_A^i(\mathbf{N}) = L_A^i(\mathbf{N})\mathbf{A} + [\mathbf{A}, L_A^i(\mathbf{N})] = L_A^i(\mathbf{N})\mathbf{A} + L_A^{i+1}(\mathbf{N}).$$

Then it is seen that

$$\begin{aligned}
\mathbf{A}^n \mathbf{N} = \mathbf{A}(\mathbf{A}^{n-1}\mathbf{N}) &= \sum_{i=0}^{n-1} \binom{n-1}{i} \mathbf{A} L_A^i(\mathbf{N}) \mathbf{A}^{n-1-i} \\
&= \sum_{i=0}^{n-1} \binom{n-1}{i} L_A^i(\mathbf{N}) \mathbf{A}^{n-i} \\
&\quad + \sum_{i=0}^{n-1} \binom{n-1}{i} L_A^{i+1}(\mathbf{N}) \mathbf{A}^{n-1-i} \\
&= \sum_{i=0}^{n-1} \binom{n-1}{i} L_A^i(\mathbf{N}) \mathbf{A}^{n-i} \\
&\quad + \sum_{i=1}^{n} \binom{n-1}{i-1} L_A^i(\mathbf{N}) \mathbf{A}^{n-i} \\
&= L_A^0(\mathbf{N})\mathbf{A}^n + \sum_{i=1}^{n-1} \left[\binom{n-1}{i} \right. \\
&\quad \left. + \binom{n-1}{i-1} \right] L_A^i(\mathbf{N}) \mathbf{A}^{n-i} + L_A^n(\mathbf{N})\mathbf{A}^0 \\
&= \sum_{i=0}^{n} \binom{n}{i} L_A^i(\mathbf{N})\, \mathbf{A}^{n-i}.
\end{aligned}$$

Proposition 2. The Baker–Campbell–Hausdorff formula

$$\operatorname{Exp}(\mathbf{A}t)\mathbf{N} = \left[\sum_{i=0}^{\infty} \frac{1}{i!} L_A^i(\mathbf{N}) t^i\right] \operatorname{Exp}(\mathbf{A}t). \tag{A.4}$$

Proof:

$$\begin{aligned}
\operatorname{Exp}(\mathbf{A}t)\mathbf{N} &= \left(\sum_{n=0}^{\infty} \frac{1}{n!} \mathbf{A}^n t^n\right)\mathbf{N} \\
&= \sum_{n=0}^{\infty} \sum_{i=0}^{n} \frac{1}{n!} \binom{n}{i} L_A^i(\mathbf{N}) t^i \mathbf{A}^{n-i} t^{n-i} \\
&= \sum_{n=0}^{\infty} \sum_{n-i=0}^{\infty} \frac{1}{i!\,(n-i)!} L_A^i(\mathbf{N}) t^i \mathbf{A}^{n-i} t^{n-i} \\
&= \sum_{n=0}^{\infty} \frac{1}{i!} L_A^i(\mathbf{N}) t^i \operatorname{Exp}(\mathbf{A}t).
\end{aligned}$$

A.2 BILINEAR EXAMPLES

To simplify the algebra, the homogeneous bilinear system BLS is considered here. It should be noted, however, that by a simple addendment of the state vector with $x_0 = 1$, the nonhomogeneous BLS may be transformed to the homogeneous case. Thus consider the homogeneous BLS

$$\frac{d\mathbf{x}}{dt} = [\mathbf{A} + \mathbf{N}u]\mathbf{x}, \tag{A.5}$$

where $\mathbf{x} \in \mathbb{R}^n$, $u \in \mathbb{R}$; $\mathbf{N}$ is an $n \times n$ constant real matrix, and $\mathbf{x}(0) = \mathbf{x}_0$.

Suppose that $\mathbf{A}$, $\mathbf{N}$, $L_A^1(\mathbf{N}) = [\mathbf{A}, \mathbf{N}]$, $L_A^2(\mathbf{N})$, . . . , form a Lie algebra of dimension $m \leq n^2$. Let $\mathbf{E}_1, \mathbf{E}_2, \ldots, \mathbf{E}_m$ form a basis set such that it is assumed that

$$\mathbf{x}(t) = \bigcap_{i=1}^{m} \exp\,[\mathbf{E}_i f_i(t)]\mathbf{x}_0. \tag{A.6}$$

Here $f_i(0) = 0$ and $f_i(t)$ are scalar functions which sometimes may be obtained by substitution. Subsequently, differentiation of (A.6) yields

$$\frac{d\mathbf{x}}{dt} = \sum_{j=1}^{m} \left[\bigcap_{i=1}^{j-1} \exp\,(\mathbf{E}_i f_i)\right] \mathbf{E}_j \left[\bigcap_{i=j}^{m} \exp\,(\mathbf{E}_i f_i)\right] \left(\frac{df_j}{dt}\right) \mathbf{x}_0. \tag{A.7}$$

Equation (A.7), in turn, is substituted, along with (A.6), into (A.5). Unfortunately, $f_i(t)$, $i = 1, \ldots, m$, and hence $\mathbf{x}(t)$ are not always solvable this way. It can be shown that BLS (A.5) is solvable in this way if there exists a nonsingular matrix $\mathbf{P}$ such that $\mathbf{PAP}^{-1}$ and $\mathbf{PNP}^{-1}$ are triangular matrices. Two special cases are presented here.

Example A.1

For any $\mathbf{A}$, if $[\mathbf{A}, \mathbf{N}] = c\mathbf{N}$, where c is a constant scalar, choose $\mathbf{E}_1 = \mathbf{A}$, $\mathbf{E}_2 = \mathbf{N}$. Then, from (A.4),

$$\begin{aligned}\text{Exp}\,(\mathbf{A}f)\mathbf{N}\,\text{Exp}\,(-\mathbf{A}f) &= \mathbf{N} + [\mathbf{A}f, \mathbf{N}] + 1/2![\mathbf{A}f, [\mathbf{A}f, \mathbf{N}]] \\ &\quad + 1/3![\mathbf{A}f, [\mathbf{A}f, [\mathbf{A}f, \mathbf{N}]]] + \cdots \\ &= \mathbf{N} + [cf + 1/2!(cf)^2 + 1/3!(cf)^3 + \cdots]\mathbf{N} \\ &= e^{cf}\mathbf{N}.\end{aligned}$$

Assume from (A.6) that

$$\mathbf{x}(t) = \text{Exp}\,(\mathbf{A}f_1)\,\text{Exp}\,(\mathbf{N}f_2)\mathbf{x}_0;$$

then, from (A.5),

$$\begin{aligned}\frac{d\mathbf{x}}{dt} &= \mathbf{A}\,\text{Exp}\,(\mathbf{A}f_1)\,\text{Exp}\,(\mathbf{N}f_2)\mathbf{x}_0\frac{df_1}{dt} \\ &\quad + \text{Exp}\,(\mathbf{A}f_1)\mathbf{N}\,\text{Exp}\,(\mathbf{N}f_2)\mathbf{x}_0\frac{df_2}{dt} \\ &= \left(\mathbf{A}\frac{df_1}{dt} + e^{cf}\mathbf{N}\frac{df_2}{dt}\right)\mathbf{x},\end{aligned}$$

so

$$\frac{df_1}{dt} = 1, \qquad e^{cf}\frac{df_2}{dt} = u(t),$$

and

$$\begin{aligned}f_1 &= t, \qquad f_2 = \int_0^t e^{-c\tau}u(\tau)\,d\tau, \\ \mathbf{x}(t) &= \text{Exp}\,(\mathbf{A}t)\,\text{Exp}\left[\mathbf{N}\int_0^t e^{-c\tau}u(\tau)\,d\tau\right]\mathbf{x}_0.\end{aligned} \tag{A.8}$$

Recall that (A.8) also is given by (5.12).

Example A.2

Suppose that for any $\mathbf{A}$, $\mathbf{N}$, $[\mathbf{A}, \mathbf{N}] = \mathbf{B}$, $[\mathbf{A}, \mathbf{B}] = \mathbf{0}$, $[\mathbf{N}, \mathbf{B}] = \mathbf{0}$. Then

$$\begin{aligned} \mathbf{AB} &= \mathbf{BA}, \qquad \mathbf{NB} = \mathbf{BN}, \qquad \mathbf{AN} = \mathbf{NA} + \mathbf{B}, \\ \mathbf{A}^2\mathbf{N} &= \mathbf{ANA} + \mathbf{AB} = \mathbf{NA}^2 + \mathbf{BA} + \mathbf{AB} = \mathbf{NA}^2 + 2\mathbf{BA}, \\ &\vdots \qquad\qquad\quad \vdots \\ \mathbf{A}^n\mathbf{N} &= \mathbf{NA}^n + n\mathbf{BA}^{n-1}, \end{aligned}$$

and

$$\begin{aligned} \text{Exp}\,(\mathbf{A}t)\mathbf{N} &= \left(\sum_{n=0}^{\infty} \frac{1}{n!}\mathbf{A}^n t^n\right)\mathbf{N} \\ &= \mathbf{N}\left(\sum_{n=0}^{\infty} \frac{1}{n!}\mathbf{A}^n t^n\right) + \mathbf{B}\left(\sum_{n=0}^{\infty} \frac{1}{n-1!}\mathbf{A}^{n-1} t^n\right) \\ &= (\mathbf{N} + \mathbf{B}t)\,\text{Exp}\,(\mathbf{A}t). \\ \text{Exp}\,(\mathbf{A}t)\mathbf{N}^2 &= (\mathbf{N} + \mathbf{B}t)^2\,\text{Exp}\,(\mathbf{A}t). \end{aligned}$$

Choose $\mathbf{E}_1 = \mathbf{A}$, $\mathbf{E}_2 = \mathbf{N}$, $\mathbf{E}_3 = \mathbf{B}$, and assume that

$$\mathbf{x} = \text{Exp}\,(\mathbf{A}f_1)\,\text{Exp}\,(\mathbf{N}f_2)\,\text{Exp}\,(\mathbf{A}f_3)\mathbf{x}_0, \qquad f_i = 0, \qquad i = 1, 2, 3.$$

Then

$$\begin{aligned} \frac{d\mathbf{x}}{dt} &= \mathbf{A}\,\text{Exp}\,(\mathbf{A}f_1)\,\text{Exp}\,(\mathbf{N}f_2)\,\text{Exp}\,(\mathbf{B}f_3)\mathbf{x}_0\frac{df_1}{dt} \\ &\quad + \text{Exp}\,(\mathbf{A}f_1)\mathbf{N}\,\text{Exp}\,(\mathbf{N}f_2)\,\text{Exp}\,(\mathbf{B}f_3)\mathbf{x}_0\frac{df_2}{dt} \\ &\quad + \text{Exp}\,(\mathbf{A}f_1)\,\text{Exp}\,(\mathbf{N}f_2)\mathbf{B}\,\text{Exp}\,(\mathbf{B}f_3)\mathbf{x}_0\frac{df_3}{dt} \\ &= \left[\mathbf{A}\frac{df_1}{dt} + (\mathbf{N} + \mathbf{B}t)\frac{df_2}{dt} + \mathbf{B}\frac{df_3}{dt}\right]\mathbf{x} \\ &= \left[\mathbf{A}\frac{df_1}{dt} + \mathbf{N}\frac{df_2}{dt} + \mathbf{B}\left(t\frac{df_2}{dt} + \frac{df_3}{dt}\right)\right]\mathbf{x}, \end{aligned}$$

so

$$\frac{df_1}{dt} = 1, \qquad \frac{df_2}{dt} = u, \qquad t\frac{df_2}{dt} + \frac{df_3}{dt} = 0;$$

then from (A.6) with

$$f_1 = t, \qquad f_2 = \int_0^t u(\tau)\,d\tau, \qquad f_3 = \int_0^t \tau u(\tau)\,d\tau,$$
$$\mathbf{x} = \text{Exp}\,(\mathbf{A}t)\,\text{Exp}\left(\mathbf{N}\int_0^t u(\tau)\,d\tau\right)\text{Exp}\left(B\int_0^t \tau u(\tau)\,d\tau\right)\mathbf{x}_0. \tag{A.9}$$

Example A.3

If $\mathbf{A}$, $\mathbf{N}$ are 2×2 triangular matrices in (A.5), we can choose

$$\mathbf{E}_1 = \mathbf{I}_2 = \begin{bmatrix} 1 & 0 \\ 0 & 1 \end{bmatrix}, \qquad \mathbf{E}_2 = \begin{bmatrix} 0 & 0 \\ 0 & 1 \end{bmatrix}, \qquad \mathbf{E}_3 = \begin{bmatrix} 0 & 0 \\ 1 & 0 \end{bmatrix};$$

then

$$[\mathbf{E}_2, \mathbf{E}_3] = \mathbf{E}_2\mathbf{E}_3 - \mathbf{E}_3\mathbf{E}_2 = \mathbf{E}_3,$$
$$[\mathbf{E}_2, [\mathbf{E}_2, \mathbf{E}_3]] = [\mathbf{E}_2, \mathbf{E}_3] = \mathbf{E}_3, \ldots,$$
$$\text{Exp}\,(\mathbf{E}_2 f)\mathbf{E}_3 = e^f\mathbf{E}_3\,\text{Exp}\,(\mathbf{E}_2 f),$$
$$[\mathbf{E}_1, \mathbf{E}_2] = 0, \qquad [\mathbf{E}_1, \mathbf{E}_3] = 0,$$
$$\text{Exp}\,(\mathbf{E}_1 f)\mathbf{E}_2 = \mathbf{E}_2\,\text{Exp}\,(\mathbf{E}_1 f), \qquad \text{Exp}\,(\mathbf{E}_1 f)\mathbf{E}_3 = \mathbf{E}_3\,\text{Exp}\,(\mathbf{E}_1 f).$$

Let $\mathbf{y}(t) = \mathbf{P}\mathbf{x}(t)$; then

$$\frac{d\mathbf{y}}{dt} = \mathbf{P}\frac{d\mathbf{x}}{dt} = (\mathbf{PAP}^{-1} + \mathbf{PNP}^{-1}u)(\mathbf{Px}) = (\mathbf{A} + \mathbf{N}u)\mathbf{y}.$$

Assume that

$$\mathbf{PAP}^{-1} = \mathbf{A} = \begin{bmatrix} a_1 & 0 \\ a_3 & a_2 \end{bmatrix}, \qquad \mathbf{PNP}^{-1} = \mathbf{N} = \begin{bmatrix} n_1 & 0 \\ n_3 & n_2 \end{bmatrix},$$

and

$$\mathbf{y} = \text{Exp}\,(\mathbf{E}_1 f_1)\,\text{Exp}\,(\mathbf{E}_2 f_2)\,\text{Exp}\,(\mathbf{E}_3 f_3)\mathbf{y}_0;$$

then

$$\frac{d\mathbf{y}}{dt} = \mathbf{E}_1\,\text{Exp}\,(\mathbf{E}_1 f_1)\,\text{Exp}\,(\mathbf{E}_2 f_2)\,\text{Exp}\,(\mathbf{E}_3 f_3)\mathbf{y}_0\frac{df_1}{dt}$$
$$+ \text{Exp}\,(\mathbf{E}_1 f_1)\mathbf{E}_2\,\text{Exp}\,(\mathbf{E}_2 f_2)\,\text{Exp}\,(\mathbf{E}_3 f_3)\mathbf{y}_0\frac{df_2}{dt}$$

$$+ \operatorname{Exp}(\mathbf{E}_1 f_1) \operatorname{Exp}(\mathbf{E}_2 f_2)\mathbf{E}_3 \operatorname{Exp}(\mathbf{E}_3 f_3)\mathbf{y}_0 \frac{df_3}{dt}$$

$$= \left[\mathbf{E}_1 \frac{df_1}{dt} + \mathbf{E}_2 \frac{df_2}{dt} + \mathbf{E}_3 \exp(f_2) \frac{df_3}{dt}\right]\mathbf{y}.$$

Then, comparing like terms,

$$f_1 = a_1 t + n_1 \int_0^t u(\tau)\, d\tau,$$

$$f_2 = (a_2 - a_1)t + (n_2 - n_1) \int_0^t u(\tau)\, d\tau,$$

$$f_3 = a_3 \int_0^t \exp[(a_1 - a_2)\tau] \exp[(n_1 - n_2) \int_0^t u(\sigma)\, d\sigma]\, d\tau$$

$$+ n_3 \int_0^t u(\tau) \exp[(a_1 - a_2)\tau] \exp[(n_1 - n_2)\tau \int_0^t u(\sigma)\, d\sigma]\, d\tau.$$

Since

$$\operatorname{Exp}(\mathbf{E}_1 f_1) = \exp(f_1)\mathbf{I}_2,$$

$$\operatorname{Exp}(\mathbf{E}_2 f_2) = \begin{bmatrix} 1 & 0 \\ 0 & \exp(f_2) \end{bmatrix}, \qquad \operatorname{Exp}(\mathbf{E}_3 f_3) = \begin{bmatrix} 1 & 0 \\ f_3 & 1 \end{bmatrix},$$

it follows that

$$\operatorname{Exp}(\mathbf{E}_1 f_1) \operatorname{Exp}(\mathbf{E}_2 f_2) = \begin{bmatrix} \exp(f_1) & 0 \\ 0 & \exp(f_1 + f_2) \end{bmatrix}$$

$$= \begin{bmatrix} \exp(a_1 t + n_1 \int_0^t u(\tau)\, d\tau) & 0 \\ 0 & \exp(a_2 t + n_2 \int_0^t u(\tau)\, d\tau) \end{bmatrix}.$$

Hence

$$\mathbf{x}(t) = \mathbf{P}^{-1}\mathbf{y}(t) = \mathbf{P}^{-1} \operatorname{Exp}(\mathbf{E}_1 f_1) \operatorname{Exp}(\mathbf{E}_2 f_2) \operatorname{Exp}(\mathbf{E}_3 f_3)\mathbf{P}(\mathbf{P}^{-1}\mathbf{y}_0)$$

$$= \mathbf{P}^{-1} \begin{bmatrix} \exp(a_1 t + n_1 \int_0^t u(\tau)\, d\tau) & 0 \\ f_3 \exp(a_2 t + n_2 \int_0^t u(\tau)\, d\tau) & \exp(a_2 t + n_2 \int_0^t u(\tau)\, d\tau) \end{bmatrix} \mathbf{P}\mathbf{x}_0. \qquad \text{(A.10)}$$

REFERENCES

[A1] Åstrom, K., "On a First-Order Stochastic Differential Equation," *Int. J. Control., 1,* 301–326, 1965.

[A2] Aoki, M., "Some Examples of Dynamic Bilinear Models in Economics," in [R2], 163–169.

[A3] Anderson, B. D. O., and Moore, J., *Linear Optimal Control Theory,* Prentice-Hall, Englewood Cliffs, N.J., 1971.

[A4] Athans, M., and Falb, P. L., *Optimal Control,* McGraw-Hill, New York, 1966.

[A5] Aseltine, J. A., Mancini, A. R., and Sarture, C. W., "A Survey of Adaptive Control Systems," *IRE Trans. Autom. Control, PGAC-5,* 102–108, 1958.

[A6] Åstrom, K. J., Anton, J. J., and Arzen, K. E., "Expert Control," *Automatica, 22,* 277–286, 1986.

[A7] Anderson, B. D. O., et al., *Stability of Adaptive Systems: Passivity and Averaging Analysis,* MIT Press, Cambridge, Mass., 1986.

[A8] Åstrom, K. J., and Wittenmark, B., *Adaptive Control,* Addison-Wesley, Reading, Mass., 1989.

[A9] Amelyanov, S. V., *Theory of Variable Structure Systems,* Science Press, Moscow, 1970.

[B1] Brockett, R. W., "On the Algebraic Structure of Bilinear Systems," in [M3], 153–168.

[B2] Balakrishnan, A. V., "On the Controllability of Nonlinear Systems," *Proc. Natl. Acad. Sci., USA, 55,* 465–468, 1966.

[B3] Brockett, R. W., "System Theory on Group Manifolds and Coset Spaces," *SIAM J. Control Optim., 10,* 265–284, 1972.

[B4] Brockett, R. W., "Volterra Series and Geometric Control Theory," *Automatica, 12,* 167–176, 1976.

[B5] Bruni, C., DiPillo, G., and Koch, G., "Bilinear Systems: An Appealing Class of Nearly Linear Systems in Theory and Applications," *IEEE Trans. Autom. Control, AC-19,* 334–348, 1974.

[B6] Bellini, A., Figalli, G., and Ulivi, G., "Realization of a Bilinear Observer of the Induction Machine," *Proc. 2nd International Conference on Electronic Variable Speed Drives,* London, 175–178, 1979.

[B7] Baheti, R. S., and Mohler, R. R., "Experimental Results in the Modeling and Control of a Small Furnace," *J. Dyn. Sys. Meas. and Control, 103,* 370–374, 1981.

[B8] Bell, G. I., Perelson, A. S., and Pimbley, G. H., Jr., eds., *Theoretical Immunology,* Marcel Dekker, New York, 1978.

[B9] Banks, S. P., *Mathematical Theories of Nonlinear Systems,* Prentice-Hall, Englewood Cliffs, N.J., 1988.

[B10] Boothby, W. M., and Wilson, E. N., "Determination of the Transitivity of Bilinear Systems," *SIAM J. Control Optim., 17,* 212–221, 1979.

[B11] Brockett, R. W., "Nonlinear Systems and Differential Geometry," *Proc. IEEE, 64,* 61–72, 1978.

[B12] Baheti, R., Mohler, R., and Spang, A., "A New Cross Correlation Algorithm for Volterra Kernel Estimation of Bilinear Systems," *IEEE Trans. Autom. Control, AC-24,* 661–664, 1979.

[B13] Bryson, A., and Ho, Y., *Applied Optimal Control,* Wiley, New York, 1975.

[B14] Boltyanskii, V. G., "Sufficient Conditions for Optimality and the Justification of the Dynamic Programming Method," *SIAM J. Control Optim. 4,* 326–361, 1966.

[B15] Balcolm, D., Demuth, H. B., and Gyftopoulos, E. P., "A Cross Correlation Method for Measuring the Impulse Response of Reactor Systems," *Nucl. Sci. Eng., 11,* 159–166, 1961.

[B16] Baheti, R., Mohler, R., and Spang, A., "A New Cross Correlation Algorithm for Volterra Kernel Estimation of Bilinear Systems," *IEEE Trans. Autom. Control, AC-24,* 661–664, 1979.

[C1] Caratheodory, C., *Calculus of Variations and Partial Differential Equations of the First Order,* B. G. Teubner, Berlin, 1935, Engl. Transl., Holden-Day, San Francisco, 1967.

[C2] Chow, W. L., "Uber Systeme von linearen partiellen Differentialgleichungen erster Ordnung," *Math Ann., 117,* 98–105, 1939.

[C3] Chitke, S. D., and Lo, J. T., "Optimal Filters for Bilinear Systems with Nilpotent Lie Algebras," *IEEE Trans. Autom. Control, AC-24,* 312–327, 1978.

[C4] Cooper, L. N., "Theory of an Immune System Retrovirus," *Proc. Natl. Acad. Sci. USA, 83,* 9159–9163, 1983.

[C5] Chevalley, C., *Theory of Lie Groups,* Princeton University Press, Princeton, N.J. 1946.

[C6] Casti, J. L., *Nonlinear System Theory,* Academic Press, New York, 1985.

[C7] Coddington, E. A. and Levinson, N., *Theory of Ordinary Differential Equations,* McGraw-Hill, New York, 1955.

[D1] Derese, I. A., and Noldus, E. J., "Existence of Bilinear State Observers for Bilinear Systems," *IEEE Trans. Autom. Control, AC-26,* 590–592, 1981.

[D2] d'Allessandro, P., Isidori, A., and Ruberti, A., "Realization and Structure Theory of Bilinear Dynamical Systems," *SIAM J. Control Optim., 12,* 517–535, 1974.

[D3] Distefano, J. J., III, "A Model of the Regulation of Circulating Thyroxin Unbound and Bound to Plasma Proteins and Its Response," *Math. Biosci., 4,* 137–152, 1969.

[D4] d'Allessandro, P., "Bilinearity and Sensitivity in Macroeconomy," in [R2], 170–199.

[D5] Derese, I., and Noldus, E., "Design of Linear Feedback Laws for Bilinear Systems," *Int. J. Control, 31,* 219–237, 1980.

[D6] DiBenedetto, M. D., and Isidori, A., "Triangular Canonical Forms for Bilinear Systems," *IEEE Trans. Autom. Control, AC-23,* 877–880, 1978.

[D7] DeCarlo, R. A., Zak, S. H., and Matthews, G. P., "Variable Structure Control of Nonlinear Multivariable Systems: A Tutorial," *Proc. IEEE, 76,* 212–232, 1988.

[E1] Espana, M., and Landau, I. D., "Reduced Order Models for Distillation Columns," *Automatica, 14,* 345–357, 1978.

[E2] Elliott, D. L., "Mathematical Models of Blood Coagulation Kinetics," in [M3], 118–122.

[E3] Eisen, H., *Immunology,* Harper & Row, New York, 1974.

[E4] Eykhoff, P., *System Identification,* J. Wiley, New York, 1979.

[F1] Fliess, M., and Normand-Cyrot, D., "On the Approximation of Nonlinear Systems by Some Simple State Space Models," *Proc. 6th IFAC Symposium on Identification and System Parameter Estimation,* Arlington, Va., 1982.

[F2] Fine, N. J., "On the Walsh Functions," *Trans. Am. Math. Soc., 65,* 372–414, 1949.

[F3] Fu, K. S., Gonzalez, R. C., and Lee, C. S. G., *Robotics,* McGraw-Hill, New York, 1987.

[G1] Gutman, P. O., "Stabilizing Controllers for Bilinear Systems," *IEEE Trans. Autom. Control, AC-26,* 917–922, 1981.

[G2] Grodins, F. S., *Control Theory and Biological Systems,* Columbia University Press, New York, 1963.

[G3] Grodins, F. S., and James, G., "Mathematical Models in Respiratory Regulation," *Ann. N. Y. Acad. Sci., 109,* 852–863, 1963.

[G4] Gershwin, S. B., and Jacobson, D. H., "A Controllability Theory for Nonlinear Systems," *IEEE Trans. Autom. Control, AC-16,* 37–46, 1971.

[G5] Gangaas, D., Bruce, K. R., Blight, J. D., and Ly, V. L., "Application of Modern Synthesis to Aircraft Control: Three Case Studies," *IEEE Trans. Autom. Control, AC-31,* 995–1014, 1986.

[H1] Hermes, H., "Controllability and the Singular Problem," *SIAM J. Control Optim., 3,* 241–266, 1965.

[H2] Hofer, E. P., and Tibken, B., "An Iterative Method for the Finite Time Bilinear Quadratic Control Problem," *J. Optim. Theory Appl.*, *57* (3), 1988.

[H3] Hermann, R., "On the Accessibility Problem in Control Theory," *National Symposium on Nonlinear Differential Equations and Nonlinear Mechanics,* (J. La Salle and S. Lefschetz, eds.) 328–332, Academic Press, New York, 1963.

[H4] Haynes, G. W., and Hermes, H., "Nonlinear Controllability via Lie Theory," *SIAM J. Control Optim.*, *8,* 450–460, 1970.

[H5] Hermann, R., and Krener, A. J., "Nonlinear Controllability and Observability," *IEEE Trans. Autom. Control, AC-22,* 728–740, 1977.

[H6] Hara, S., and Furuta, K., "Minimal Order State Observer for Bilinear Systems," *Int. J. Control, 24,* 705–718, 1976.

[H7] Hofer, E. P., and Tibken, B., "Systematic Observer Design for Bilinear Systems," *Proc. IEEE International Symposium on Circuits and Systems,* Portland, Oreg., 1611–1616, 1989.

[H8] Haussmann, V., "Optimal Stationary Control with Control and State Dependent Noise," *SIAM J. Control Optim.*, *9,* 184–198, 1971.

[H9] Hoffmann, G. W., and Hraba, T., eds., *Immunology and Epidemiology,* Springer-Verlag, New York, 1986.

[H10] Harmuth, H. F., *Transmission of Information by Orthogonal Functions,* Springer-Verlag, New York, 1972.

[I1] Isidori, A., "Direct Construction of Minimal Bilinear Realizations from Nonlinear Input–Output Maps," *IEEE Trans. Autom. Control, AC-18,* 626–631, 1973.

[J1] Jacobsen, D. H., *Extensions of Linear-Quadratic Control, Optimization and Matrix Theory,* Academic Press, London, 1979.

[K1] Kolodziej, W. J., and Mohler, R. R., "State Estimation and Control of Conditionally Linear Systems," *SIAM J. Control Optim.*, *24,* 497–508, 1986.

[K2] Krener, A. J., "Bilinear and Nonlinear Realizations of Input/Output Maps," *SIAM J. Control Optim.*, *13,* 827–834, 1975.

[K3] Kucera, J., "Solution in Large of Control Problem $\dot{x} = [A(1 - u) + Bu]x$," *Czech. Math. J.*, *16,* 600–622, 1966.

[K4] Koenig, H., "Variable Parameter Structures in Technology Assessment and Land Use," in [R2], 200–205.

[K5] Keepin, G. R., *Physics of Nuclear Kinetics,* Addison-Wesley, Reading, Mass., 1965.

[K6] Koditschek, D. E., and Narendra, K. S., "Stabilizability of 2nd Order Bilinear Systems," *IEEE Trans. Autom. Control, AC-28,* 987, 1983.

[K7] Kucera, J., "On Accessibility of Bilinear Systems," *Czech. Math. J.*, *20,* 160–168, 1970.

[K8] Kostyukovskii, Y. M., "Observability of Nonlinear Controlled Systems," *Autom. Remote Control, 9,* 1384–1396, 1968.

[K9] Kou, S. R., Elliott, D. L., and Tarn, T. J.,. "Observability of Nonlinear Systems," *Inf. Control, 22,* 89–94, 1973.

[K10] Karanam, V. R., Frick, P. A., and Mohler, R. R., "Bilinear System Identification by Walsh Functions," *IEEE Trans. Autom. Control, AC-23,* 709–713, 1978.

[K11] Kirk, D., *Optimal Control Theory: An Introduction,* Prentice-Hall, Englewood Cliffs, N.J., 1970.

[K12] Kelley, H. J., "A Transformation Approach to Singular Subarcs in Optimal Trajectory and Control Problems," *SIAM J. Control Optim., 2,* 234–240, 1965.

[K13] Karnopp, D., "Active Damping in Road Vehicle Suspension Sysems," *Vehicle Sys. Dyn., 12,* 291–316, 1983.

[K14] Krasnicki, E., "Comparison of Analytic and Experimental Results for a Semi-active Vibrator Isolator," *Proc. 50th Shock and Vibration Symposium,* Oct. 1979.

[K15] Kimbrough, S., "Bilinear Modelling and Regulation of Variable-Component Suspensions," personal communication, 1986.

[K16] Kresselmeir, G., and Anderson, B. D. O., "Robust Model Reference Adaptive Control," *IEEE Trans. Autom. Control, AC-31,* 127–133, 1986.

[K17] Koivo, A. J., and Guo, T. H., "Adaptive Linear Controller for Robotic Manipulators," *IEEE Trans. Autom. Control, AC-28,* 162–171, 1983.

[L1] Lewis, J. B., "The Use of Nonlinear Feedback to Improve the Transient Response of a Servomechanism," *Trans. AIEE, 71,* 449–453, 1953.

[L2] Landau, I. D., "On the Optimal Regulator Problem and Stabilization of Bilinear Systems," *Rep. Lab. d'Automatique (CNRS),* Grenoble, France, 1979.

[L3] Longchamp, R., "Stable Feedback Control of Bilinear Systems," *IEEE Trans. Autom. Control, AC-25,* 302–306, 1980.

[L4] Lee, K. Y., "Optimal Bilinear Control Theory Applied to Pest Management," in [M11], 173–188.

[L5] Lee, E. B., and Markus, L., "Optimal Control for Optimal Processes," *Arch. Rational Mech. Anal.,. 8,* 36–58, 1961.

[L6] Luenberger, D., "Observers for Multivariable Systems," *IEEE Trans. Autom. Control, AC-11,* 190–197, 1966.

[L7] Lee, E. B., and Markus, L., *Foundations of Optimal Control Theory*, Wiley, New York, 1967.

[L8] Lee, C. S. G., and Chung, M. J., "An Adaptive Control Strategy for Mechanical Manipulators," *IEEE Trans. Autom. Control, AC-29*, 837–840, 1984.

[M1] Mohler, R. R., *Nonlinear Systems;* Vol. 1, *Dynamics and Control*, Prentice Hall, Englewood Cliffs, N.J., 1991.

[M2] Mohler, R. R., *Bilinear Control Processes*, Academic Press, New York, 1973.

[M3] Mohler, R. R., and Ruberti, A., eds., *Theory and Applications of Variable Structure Systems*, Acadmic Press, New York, 1972.

[M4] Mohler, R. R., and Kolodziej, W. J., "Optimal Control of a Class of Nonlinear Stochastic Systems," *IEEE Trans. Autom. Control, AC-26*, 1048–1053, 1981.

[M5] Mohler, R. R., "Stability and Control of Nuclear Rocket Propulsion," *IRE Trans. Autom. Control, AC-7*, 86–96, 1962.

[M6] Mohler, R. R., and Perry, J. E., "Nuclear Rocket Engine Control," *Nucleonics, 19,* 80–84, 1961.

[M7] Mohler, R. R., "Optimal Control of Nuclear Reactors and Bilinear Systems," *Los Alamos Scientific Laboratory Report*, 1963.

[M8] Mohler, R. R., and Shen, C. N., *Optimal Control of Nuclear Reactors,* Academic Press, New York, 1972.

[M9] Mohler, R. R., and Rink, R. E., "Control with a Multiplicative Mode," *J. Basic. Eng. (ASME Trans.), 91*, 201–206, 1969.

[M10] Mohler, R. R., and Rink, R. E., "Reachable Zones for Equicontinuous Bilinear Control Processes," *Int. J. Control, 14,* 331–339, 1971.

[M11] Mohler, R. R., and Ruberti, A., eds., *Recent Developments on Variable Structure Systems, Biology and Economics*, Springer-Verlag, New York, 1978.

[M12] Marcus, S., and Willsky, A., "Algebraic Structure and Finite Dimensional Nonlinear Estimation," *SIAM J. Math. Anal., 9,* 312–327, 1978.

[M13] Mohler, R. R., "Bilinear Systems and Control," *Encyclopedia of Physical Science and Technology*, Academic Press, New York, 1987.

[M14] Mohler, R. R., Bruni, C., and Gandolfi, A., "A Systems Approach to Immunology," *Proc. IEEE, 68*, 964–990.

[M15] Mohler, R. R., Barton, C. F., and Hsu, C. S., "T and B Cells in the Immune System," in [B8], 415–436.

[M16] Marchuk, G. I., *Mathematical Models in Immunology*, Optimization Software Publication, Springer-Verlag, New York, 1983.

[M17] Merrill, S. J., "AIDS: Background and the Dynamics of the Decline of Immunocompetence," in [P2].

[M18] Mohler, R. R., and Frick, P. A., "Bilinear Demographic Control Processes," *Int. J. Policy Anal. Inf. Sys., 2,* 57–70, 1979.

[M19] Moylan, P. J., and Anderson, B. D. O., "Nonlinear Regulator Theory," *IEEE Trans. Autom. Control, AC-18,* 460–465, 1973.

[M20] Mohler, R. R., and Hwang, C. S., "Nonlinear Data Observability and Non-Gaussian Information Structures," in *Stochastic Processes in Underwater Acoustics,* (C. R. Baker, ed.), Springer-Verlag, New York, 1986.

[N1] Narendra, K. S., ed., *Adaptive and Learning Systems,* Plenum Press, New York, 1986.

[01] Oster, G., "Bilinear Models in Ecology," in [M11], 260–271.

[02] Ohta, M., and Takao, T., "A Measurement-Based Simulation for Commercial Vehicles," *Soc. Autom. Eng. Trans.*, 851217, 1985.

[P1] Patten, B. C., "Variable Structure Aspects of Ecosystems," in [M3], 213–232.

[P2] Perelson, A. S., ed., *Theoretical Immunology, Vols. 1 and 2,* Addison-Wesley, Reading, Mass., 1988.

[P3] Palais, R. S., "Natural Operation of Differential Forms," *Trans. Am. Math. Soc., 92,* 125–141, 1959.

[P4] Pontryagin, L. S., Boltyanski, V. G., Gamdrelidze, R. V., and Mischenko, E. F., *The Mathematical Theory of Optimal Processes,* (L. Neustadt, transl.), Wiley, New York, 1962.

[P5] Prather, E. E., "Passenger Car Brake System Performance Requirements," *Soc., Autom. Eng. Trans.,* 6800176, 1968.

[P6] Papoulis, A., *Probability, Random Variables and Stochastic Processes*, McGraw-Hill, New York, 1965.

[Q1] Quinn, J. P., "Stabilization of Bilinear Systems by Quadratic Feedback Controls," *J. Math. Anal. Appl., 75,* 66–80, 1980.

[R1] Rink, R. E., and Mohler, R. R., "Completely Controllable Bilinear Systems," *SIAM J. Control Optim., 6,* 477–486, 1968.

√[R2] Ruberti, A., and Mohler, R. R., eds., *Variable Structure Systems with Application to Biology and Economics,* Springer-Verlag, New York, 1975.

[R3] Rink, R. E., and Le, H. Q., "Multivariable Control of Bilinear Processes in HVAC Systems," *Proc. 24th IEEE Midwest Symposium on Circuits and Systems,* Albuquerque, N. Mex., 777–781, 1981.

[R4] Roitt, I., *Essential Immunology,* Blackwell Scientific, Oxford, 1974.

[R5] Rugh, W. J., *Nonlinear System Theory: The Volterra/Wiener Approach,* Johns Hopkins University Press, Baltimore, 1981.

[R6] Ryan, E. P., "Optimal Feedback Control of Bilinear Systems," *J. Optim. Theory Appl., 22,* 333–362, 1984.

[S1] Sussmann, H. J., "Semigroup Representations, Bilinear Approximation of Input–Output Maps and Generalized Inputs," in *Mathematical System Theory* (G. Marchesni and S. Mitter, eds.), Springer-Verlag, New York, 1976.

[S2] Slemrod, M., "Stabilization of Bilinear Control Systems with Application to Nonconservative Problems in Elasticity," *SIAM J. Control Optim., 16,* 131–141, 1978.

[S3] Sandor, J., and Williamson, D., "Design of Nonlinear Regulators for Linear Plants," *IEEE Trans. Autom. Control, AC-22,* 47–50, 1977.

[S4] Sussmann, H. J., and Jurdjevic, V. J., "Controllability of Nonlinear Systems," *J. Differential Equations, 12,* 95–116, 1972.

[S5] Schetzen, M., *The Volterra and Wiener Theories of Nonlinear Systems,* Wiley, New York, 1980.

[S6] Sastry, S., and Bodson, M., *Adaptive Control,* Prentice-Hall, Englewood Cliffs, N.J., 1989.

[T1] Tzafestas, S., Anagnostou, K., and Pimenides, T., "Stabilizing Optimal Control of Bilinear Systems with Generalized Cost," *Optimal Control Appl. Methods, 5,* 111–117, 1984.

[T2] Tregal, R. O., "Truck and Bus Brake System Performance Requirements," *Soc. Autom. Eng. Trans.,* 680175, 1968.

[T3] Tsypkin, Y. Z., "Optimality in Adaptive Control Systems," in *Uncertainty and Control,* (J. Ackermann, ed.), 153–214, Springer-Verlag, New York, 1985.

[V1] Vars, R. C., "Public Information Activities: Evaluation with Variable Coefficient Models," in [M11], 305–317.

[V2] Vidyasagar, M., *Nonlinear Systems Analysis,* Prentice-Hall, Englewood Cliffs, N.J., 1978.

[V3] Vidyasagar, M., "New Directions of Research in Nonlinear System Theory," *Proc. IEEE, 74*, 1060–1091, 1987.

[W1] Williamson, D., "Observation of Bilinear Systems with Application to Biological Control," *Automatica, 13,* 243–254, 1977.

[Y1] Yang, X., Mohler, R. R., and Burton, R. M., "Adaptive Supoptimal Filtering of Bilinear Systems," *Int. J. Control 32,* 1990.

[Z1] Zames, G., "On the Input–Output Stability of Nonlinear Time-Varying Feedback Systems, I, II," *IEEE Trans. Autom. Control, AC-11,* 228–238, 465–476, 1964.

INDEX

A

B

C

D

E

F

G

H

W

Z